G

工程翻译:口译篇

杨艳君　周薇薇　赵筱颖　著

江西高校出版社
JIANGXI UNIVERSITIES AND COLLEGES PRESS

南昌

图书在版编目(CIP)数据

工程翻译. 口译篇／杨艳君，周薇薇，赵筱颖著. --南
昌：江西高校出版社，2024.12
ISBN 978 - 7 - 5762 - 4582 - 0

Ⅰ. ①工… Ⅱ. ①杨… ②周… ③赵… Ⅲ. ①
工程技术 - 英语 - 口译 - 教材 Ⅳ. ①TB

中国国家版本馆 CIP 数据核字(2023)第 244769 号

| 策 划 编 辑 | 陈永林 | 责 任 编 辑 | 黄 倩 |
| 装 帧 设 计 | 曾 宇 | 责 任 印 制 | 李香娇 |

出 版 发 行	江西高校出版社
社 址	江西省南昌市洪都北大道 96 号
邮 政 编 码	330046
总 编 室 电 话	0791 - 88504319
销 售 电 话	0791 - 88511423
网 址	www.juacp.com
印 刷	江西新华印刷发展集团有限公司
经 销	全国新华书店
开 本	700 mm×1000 mm 1/16
印 张	20.5
字 数	346 千字
版 次	2024 年 12 月第 1 版
印 次	2024 年 12 月第 1 次印刷
书 号	ISBN 978 - 7 - 5762 - 4582 - 0
定 价	78.00 元

赣版权登字 -07 -2023 -918

前　　言

　　2013 年,中国国家主席习近平审时度势,提出共建"一带一路"宏伟倡议,成为人类发展史上具有里程碑意义的事件。十年来,在国内外各方携手努力下,共建"一带一路"落地生根、蓬勃发展,已成为开放包容、互利互惠、合作共赢、深受欢迎的国际公共产品和国际合作平台,为全球共同发展搭平台、做增量、添动力。2022 年,习近平总书记在党的二十大报告中又指出:高质量发展是全面建设社会主义现代化国家的首要任务。我们要坚持以推动高质量发展为主题,推动共建"一带一路"高质量发展。

　　在此背景下,中国进一步加快了对外开放的步伐,越来越多的中国企业,尤其是与基础建设密切相关的工程类企业"走出去"已成为经济发展的新常态。因此,培养具有国际视野、通晓国际规则、能够参与国际事务和国际竞争的国际化人才成为高等院校对接"一带一路"倡议的教育使命。鉴于时代发展对新型口译人才和口译教学的要求,以及结合学生自身的需求和特点,编者规划编写了《工程翻译:口译篇》教材。

　　本教材是根据《翻译硕士专业学位研究生教育指导性培养方案》的要求编写的一部专业教材,旨在满足翻译硕士专业学位研究生、翻译专业本科生高阶学习和工程英语翻译专业人员的学习和工作需求。本教材以国际工程行业从业者所从事的典型工作过程为主线来编写,先易后难、循序渐进,主要包括工程项目启动、招投标、合同谈判、施工准备、材料采购与设备租赁、进度管理、质量管理、成本管理、验收与付款、项目索赔、后勤保障等 11 个主题。此外，本教材以口译技能为基础，囊括多任务处理、主旨听辨、信息分层、记忆与预测、语言重组、顺句驱动等 12 项必须掌握的口译技巧，且将技巧讲解融于单句口译、对话口译、篇章口译、词汇拓展等练习中，实现技能培养、知识传授、职业修养三位一体的教学目标。

为贯彻教育部数字化教学的理念，本教材更新编写形式，利用云资源，配套电子教案，添加音频二维码，学生可使用电脑、手机 App 辅助学习，提高教学效率，同时培养信息素养与能力。

该书得到南昌工程学院"十四五"一流学科"外国语言文学"学科建设经费资助，得益于南昌工程学院外国语学院、研究生处和教务处等部门的大力支持，受获于江西省水利水电建设集团有限公司张丽雅女士的专业指导以及"专门用途英语课程群虚拟教研室"诸位教师的宝贵建议，在此表示衷心的感谢。杨艳君执笔第二、三、六、七、八、九章（约 18 万字）；周薇薇执笔第十一、十二章（约 6.6 万字）；赵筱颖执笔第一、四、五、十章（约 10 万字）。

由于时间仓促，本书难免有不足之处，敬请批评指正！

编者

2024 年 6 月

CONTENTS

目　　录

第一章 工程口译概述

现在国际化合作非常多,国际工程项目也越来越多了。工程口译存在于国际工程合作之间,并随着工程合作的国际化越来越频繁地出现在人们的视野当中,这也对工程口译人员提出了越来越高的要求。在学习工程项目流程及工程口译技巧之前,我们先来了解什么是口译、工程口译以及作为一名工程项目的译员需要注意哪些问题。

第一节 口译的定义及特点

口译是翻译的一种形式,指将一种语言所表述的内容用另一种语言即时准确地口头表达出来。职业口译是一种服务行为,目的在于保证使用不同语言的人们交际顺利,以促进国际政治、经济、文化和科技交流。

自从有了人类,有了语言,有了不同语言群体之间的交流,就有了口译。口译作为一种国际认可的职业出现于第一次世界大战末期。1919 年的巴黎和会首次借助英、法两种语言的翻译进行谈判,从而打破了法语在国际会议和外交谈判中的垄断地位,也标志着职业口译正式登上国际舞台。

虽然口译和笔译的根本任务和性质是相同的,但口译和笔译又具有各自不同的特点和要求。它们的差异主要体现在以下几个方面:

1. 口译和笔译的工作环境不同

口译具有很强的现场性和时限性。口译员必须面对公众现场翻译,口译时除非发现重大错漏,一般不能对已翻译过的内容进行大量的更正和补充。而在笔译中,译者大多有充足的时间反复阅读原文,译完初稿后,还可以反复修改润色直到满意为止。

2. 口译和笔译的工作方法不同

口译员通常独自工作,口译时一般无法求助他人,也没有时间查阅词典或其他资料。而笔译员则可随时查阅各种参考书,反复斟酌,也可请教相关领域的专家。

3.口译和笔译的反馈形式不同

由于口译是现场的翻译活动,译员面对讲话人和听众,随时可以从他们的面部表情、情绪、手势和其他身势语中接收到反馈信息,根据现场的情况和听众的反馈及时调整自己的音量、语速、用词等。而笔译工作者在翻译过程中一般没有机会与作者和读者面对面交流,也无法根据他们的反馈进行调整或更正。

4.口译和笔译的特点不同

笔译讲究"信、达、雅",口译则强调"准、顺、快"。"准"即准确,指准确理解源语信息并即时将其译成目的语;"顺"即通顺,指译员在用目的语表达源语信息时要通顺流畅,符合语言表达规范;"快"即译员反应要快,讲话者的话音刚落,译员就要开始翻译。另外,口译质量取决于参与交际的各方(如讲话者、听众和雇主)从各个角度所认定的交际效果。

第二节 口译的类型及标准

一、口译的形式

口译可根据其工作模式、场合及内容进行不同的划分。按照口译的工作模式,口译可分为交替传译和同声传译;按照口译的场合及内容,口译可分为会议口译、耳语传译、视译、联络陪同口译、法庭口译、媒体口译以及电话口译。

1.交替传译(consecutive interpreting)。交替传译,简称交传,亦称即席口译或连续口译,是比较传统的口译模式,常用于会议、媒体采访等场合。交传时,译员须在有限的时间内将讲话的内容用译入语传递给听众,通俗地讲,译员逐段翻译讲话的信息。交传要求译员有很强的笔记能力,而且会使活动的时间增加。对听众而言,他们也必须高度集中注意力,因为讲话会不断停顿。

2.同声传译(simultaneous interpreting)。同声传译简称同传,是指译员在讲话者发言的同时,通过耳机边听边译。这种口译模式自二战以来广泛用于各种国际会议。由于脑力和体力的特殊需求,同声传译一般有两位译员在隔音的翻译箱内轮流翻译,分担任务。就时间而言,同传不占用会议的时间,并且可同时使用多种语言。

3. 会议口译（conference interpreting）。会议口译用于多语种会议的场合。它包含交传和同传两种口译模式。大会译员可根据会议的性质及环境采取不同的口译模式。

4. 耳语传译（whispering interpreting）。耳语传译是同声传译的一种形式。它需要译员用耳语的方式同步、轻声地将听到的信息传译给身边的一两位听众。耳语传译时译员无须使用同传设备，其服务对象只是少数（一般不超过三人）不懂工作语言的听众。

5. 视译（sight interpreting）。视译是一种读、译相结合的口译方式，指译员边看书面材料边将内容准确、完整地传译给听众。译员有时直接按照所给的书面材料读译，有时需边听讲话，边看书面材料，同时进行传译。这是一种比较特殊的口译形式，常见于大会口译。

6. 联络陪同口译（escort interpreting）。联络陪同口译一般指译员陪同个人或团体旅游、参观、参加会议、接受采访等。其口译模式是交传。

7. 法庭口译（court interpreting）。法庭口译在国外较为常见，尤其在一些移民国家，主要用于法庭和司法机关。在这种口译场合，交传和同传都会被使用。法庭口译要求译员公正、客观地传递信息。因此，国外的译员在法庭口译之前需宣誓。

8. 媒体口译（media interpreting）。媒体口译涵盖面甚广，包括记者招待会、公共宣传、采访、电影、录像以及电视和广播节目的口译。媒体口译场合中交传和同传都会被使用。

9. 电话口译（over-the-phone interpreting）。电话口译是指远程口译，交流双方及译员处在不同的地理位置。译员通过电话口译涉及医疗、公共服务及司法等方面的个案。这种口译场合一般采用交传。由于看不到交流的双方，译员在电话口译时有一定的困难。

二、口译的标准

关于翻译的标准，西方翻译理论家提出了"等效""等值"的原则，我国翻译家严复提出了"信、达、雅"的衡量标准。虽然翻译界对此有不同的解释，但多年来"信、达、雅"一直是衡量笔译质量的标准，也是我国广大翻译工作者从事笔译所遵循的基本原则。其实"信、达、雅"的原则也适应于口译，只是口译受工作条件和客观环境的限制,"信、达、雅"的侧重点有所不同而已。口译的性质和特点决定了口译的标准应为"准确、流利、易懂"。

准确是口译的基本标准。准确就是"信"。在口译中,"信"就是忠实地传达说话人的原意,就是口译的信度。具体地说,在内容上,译员要全面完整、准确无误地传达谈话的议题、观点,涉及的事实、细节、数字、时间、地点等,不能有任何的疏漏和差错。准确是口译的生命线,是译员工作责任心和业务能力的集中体现。由于口译内容的范围很广,有涉及立场和观点的政治会谈,有包括各种数字的经贸谈判,即使在日常翻译中也常常涉及具体的时间、地点和细节,因此,口译内容上的任何差错或失真,都可能造成政治上的原则性错误、经济上的重大损失或工作上的严重失误。口译中,译员不可只顾速度而忽视口译的准确性。任何粗心大意、敷衍了事的作风都可能造成难以挽回的后果。口译的准确性也体现在语言表达方面。在语言表达上译员要做到语音、语调正确,准确掌握词义、词性、词的变化与搭配,正确运用句型、时态、语气,做到语音、语调、词法、语法准确无误。译员语言表达上的任何错误都可能造成交际双方谈话内容上的含糊不清,似是而非,甚至引起严重误解。不准确的口译不可能真实地传达说话人的意愿,也不符合口译的准确原则。口译的准确性还体现在风格上。译员要正确运用语态、语气、情态等,再现说话人的情感、情绪和语气,真正做到传情达意,再现说话人"原汁原味"的谈话风格。

流利是口头表达的基本要求,也是口译的另一重要标准。在口译中,"达"就是语言通达、通顺、流畅。口译要做到流利,一要"快",二要"畅"。译员必须迅速、及时地把一方的话语信息传达给另一方,做到语速流畅、节奏适当、反应敏捷、出口利落,不能吞吞吐吐。

易懂是指口译的语言要口语化,要简洁明快、直截了当,符合译入语的表达方式,使人一听就懂。特别是汉译英时,一定要避免汉语式的英语,或"对号入座"式的"死译",说出的英语不能让外国听众不知所云。

当然,准确、流利、易懂的口译是建立在译员对两种交际语言的技巧和文化知识熟练掌握及快速反应基础之上的。因此,练好语言基本功,加强口译实践锻炼是译员获得良好口译效果的根本途径。

第三节　工程口译的特点

工程口译是口译的一个分支,属于技术口译的范畴,它对口译的准确性要求非常高。工程口译通常是指在大型工程项目施工的过程中,由于在设备进口、技术引进、设备安装和设备调试等方面需要与国外的技术人员、工程师或相关人员进行交流所衍生的口译服务。工程项目大多专业性强,周期较长,因此需要中外技术人员不断交流和及时沟通,以保证项目的顺利进行和完工。工程口译与普通口译一样,具有以下一些特点:

(1)口语特点。口语有别于书面语,是谈话时使用的语言;它具有即时性,说话者总是边思考、边组织、边说话;口语的存留时间短暂,给口译的研究和实践工作带来一定的困难;口语的言语发布速度快,口译实践中有的发言低于正常语速,即平均每分钟100至120字,但更多情况下发言语速超过一般语速;口语是互动性交际活动,交际双方需不断对信息及时做出反应;口语交际的各方可以即刻观察到对方的反应,以便继续或中断口语交际活动;口语有自己的语体、语级和语类。口语的这些特点会对口译产生重大影响。

(2)交际特点。口译的工作环境、工作方法、工作时间与笔译不同,其现场性和时限性决定了它是一种更为直接的交际行为。

(3)不可预测性。口译人员需要在准备有限的情况下,即刻进入双语语码切换状态,进行现场的口译操作。口译话题千变万化,往往难以预测。

(4)现场压力。口译场面有时非常严肃庄重,正式场合的严肃气氛会给经验不足的译员造成很大的心理压力。

此外,工程口译还具有区别于一般口译的特点:

(1)专业性。工程口译涉及面非常广,既包括工程项目的招投标、施工与竣工的全过程,也包括施工会议、合同谈判等各种正式和非正式场合。这就要求口译员的知识面越宽越好,不仅要懂得文化、外交、旅游等常识,更要懂得基本的工程知识和专业术语。专业性是工程口译有别于普通口译的重要特点。

(2)现场特殊性。一般口译的现场大多是会议厅或谈判桌,而工程口译的现场则十分多样化,比如建筑施工现场、设备安装调试现场、厂房参观现场等,总之一切工程活动的地方都可以是工程口译现场。我们很难要求这些地方如会议厅一样安静有序,所以工程口译员必须具备良好的环境适应能力,适应工程现场的噪声与恶劣环境等。

(3)高度的互动性。工程口译的现场性决定了其有着高度的互动性。一般的会议口译现场,译员或在台前,或在同传箱子中进行翻译。尽管一直强调互动的重要性,但实际上译员最多只能做到与观众进行眼神交流,或观察现场反应。而在工程口译中,由于现场的多样性,译员可以随时与中外双方互动交流。比如,工程现场如果译员有一个术语不知道,可以立即与讲话者沟通,讲话者可以进一步通过语言或手势解释该术语,之后译员可以译出其大致意思。有时由于专业相通,对方技术人员甚至无须翻译就明白其意思了。

专业性、现场特殊性和高度的互动性是工程口译的突出特点,了解和把握这些特点有利于译者更好地进行工程口译。

第四节　工程口译的难点

从事工程口译工作的译员都会发现口译过程中会遇到很多长难句,因而长难句的处理问题是口译译员不得不面对的难题之一;工程英语专业性强,专业术语繁多,这就给口译工作带来了高难度;再者,项目上聘请的海外专家几乎遍布各大洲,克服口音问题是一大关键;同时,客户性格迥异,所受教育程度也有所不同,采用何种方式进行沟通至关重要;此外,口译工作环境也有很多困难需要克服,如现场环境嘈杂,酷热难耐,如何抵抗高温、高压和噪声是亟待解决的问题。

一、长难句的处理问题

学界对于长句的定义没有统一的标准。笔者认为长句可分为两种,一种单纯地指字数多、句式长;另一种指句子修饰成分多、句式复杂。一般情况下,长句同时包含这两种特征。难句即复杂句,是依简单句而言的,而复杂句又包括并列句和复合句以及并列复合句。一般情况下,长句都长且句式结构复杂,具体表现为句子中添加了若干从句和其他修饰成分。长难句处理问题是口译译员需要面对的难题,原因在于,长难句给口译译员的记忆带来了一定的负担,可能会导致译文的完整性丧失。此外,英、汉两种语言不同,分析和处理长难句的方法也各不相同,这就需要口译译员在处理长难句时运用一定的技巧、经验和方法。

例 1:I agree with the lifting team that the Manitou can replace both forklift and crane to carry all the black hoses all the way back to the empty area between the white tent and the parking lot.

翻译:我同意吊装团队的意见,用曼尼通取代叉车和吊车,把所有的黑色管子转移到白色帐篷和停车场之间的空旷地带。

分析:这句话使用了主从句、动词的非谓语形式等形合手段,使得句子长而难。口译的即时性决定了译员没有太多可以推敲琢磨的时间。除去句子成分分析不说,句子中提供的细节也较多,仅名词就有"吊装团队、曼尼通、叉车、吊车、黑色管子、空旷地带、白色帐篷、停车场"至少八个,要是不依靠大脑的想象和场景勾勒,是很难记忆并翻译的。译文将源语者的意图较全面地展现了出来,但是美中不足的是,在现场翻译时将"吊装团队"这一细节落下了,译员应加强记忆,力求在细节方面做得更好。

英语的长难句主要表现在句子长、句中修饰成分多且关系结构复杂,而汉语的长难句的特征包括句子长、句式结构松散且包含多个隐形主语。由于汉语重意合,句子间缺乏连接词,句式显得不紧凑,汉译英时常常需要找到句子间的逻辑关系,将句子重新组合,并通过添加逻辑衔接词等成分将汉语短句变成英语长句。

例 2:50 吨重的吊车在一列停靠,我去把它开到三列,让卡车司机现在把卡车也开到三列,准备吊装、运输制冷器。

翻译:I will drive the 50-ton crane that is in train 1 to train 3 and also tell truck driver to move the truck to train 3 to get ready for lifting and transporting of air cooler.

分析:这句话由四个分句构成,并且还隐含着四个不同的主语,这和汉语"形散意合"的特点是一致的,翻译成英语时要找到句子之间的联系,确定合适的主语,并通过逻辑衔接词连接整个句子。这个句子大致可整合为如下结构:我去把它[50 吨重的(停在一列的)吊车]开到三列,(同时)通知开车司机把卡车也开到三列,(为的是)吊装和运输制冷器。现场的译文是"50-ton crane now is in train 1. I will drive it to train 3 and tell truck driver to move it to train 3. It's to get ready for lifting and transporting of air cooler. "。这一译文是有效的,但是如果能够更好地把握句子主干,句式结构就能够更加简洁流畅。

二、专业术语繁多的问题

"工程"是数学和科学的某种应用,这就决定了工程学科的复杂性和专业性。相应地,对于工程口译而言,翻译含有专业术语的句子成了一大难点,如果对专业术语把握不准就会影响译文的准确性。因此,译员需要掌握一定的专业知识和术语。

例1:Have the shell been cured?

翻译:壳层固化了吗?

分析:这句话看起来就是一个简单的疑问句,但是这里的"cure"却不是平时人们熟知的"治愈"之意,一句简单的话因"cure"一词的翻译变得有难度。译员第一次听到这个词是在为酸气分离器进行壳层固化时,由于刚刚接触这一领域,尚不熟悉这个词的意思。但根据现场的状况,译员对其进行模糊化处理,翻译成"壳层处理了吗?"虽不至于译错,但是却不够准确,所以译员向在场的中外双方请教,更好地解决了现场的沟通问题。之后译员为了进一步了解和掌握"cure"一词,在韦氏词典上找到了专业的解释——"prepare by drying, salting, or chemical processing in order to preserve"。在工程口译中,类似的专业词汇是极多的,译员能做的就是平时尽量多积累,现场多学习。

专业术语的口译问题不仅存在于英译汉中,同样也会出现在汉译英中。笔者在下文给出了相应的例证和分析。

例2:高压放空和低压放空都是十分危险,除施工人员外,其他人员不得靠近。

翻译:HOG and LOG are dangerous, so other people except operational guys shouldn't be around it.

分析:在天然气工程领域,大修期间会进行高压放空和低压放空,其英文全称分别为 high-pressure off-gas 和 low-pressure off-gas,现场外国专家通常将其简称为 HOG 和 LOG。笔者将其直译为"release of high pressure and low pressure",这是不准确的。这样的专业术语,都是现场翻译的一大障碍,译员必须多学习。

例3:一列脱水区域的螺纹连接是目前工作的重中之重,是进行壳层和管层压力测试的重要前提,一定要把它做好。

翻译:Threaded connection in train 1 dehydration is the priority job now, which is the precondition of conducting pressure test on both shell side and tube side.

分析:这句话中包含三个专业术语,即"螺纹连接""壳层压力测试"和"管层压力测试"。译员初次接触"螺纹连接"这一概念时还有些生疏,但为了不影响工作进程,译员尝试用解释性的话语进行口译,告诉外方负责人:"A kind of connection job in train 1 is the priority now."。对方很快便理解了,告知译员是"threaded connection",之后译员就能够使用更准确的术语来进行传译了。"壳层压力测试"和"管层压力测试"也属于专业术语,但译员此前在锅炉房接触过相关的口译工作,对这部分内容有所了解,所以在口译过程中能够准确翻译。

三、口音问题

国际工程项目上的外国专家数以百计,他们来自世界各地,所说的英语夹杂的口音也各不相同。口音问题会影响口译译员的听辨质量,进而导致译文质量下降。解决这一问题需要译员积极地适应不同的口音,了解不同口音的发音特点。

例 1:I'd like to change the color. = I'd lig do change de gala.

翻译:我想把颜色换一换。

例 2:Grinding comes first. = Glinding comes first.

翻译:先打磨。

分析:例 1 是项目承包公司的一位来自印度的工作人员所说的一句话。印度英语发音有以下特点:P 发 B,T 发 D,K 发 G,R 发 L;没有爆破音和清辅音;语音语调也有所不同。例 2 是一位苏格兰员工的英语发音,苏格兰英语中存在卷舌的颤音,音调也不同于英格兰地区,所以例句中的 grinding 听起来就像是 glinding。译员在项目上会遇到各种不同的口音,需要花费时间和精力去适应。

四、文化差异问题

参与工程项目的工作人员,其性格是不同的。不同企业文化孕育的人也有所差异,有热情宽容的现场负责人,同样也有不苟言笑,甚至在常人看来有些苛刻的现场负责人。一位译员曾经为涉外机电项目脱硫区域的现场负责人提供口译。第一天在现场召开开工前的安全工具箱会议时,该负责人非常严肃地告知工人:"If you wanna kill yourself, don't do it on my job site. Do it at home. I'm not your mother, so I won't tell you everything you should know."译员一开始感觉直译过来可能不太礼貌,但因为涉及安全问题,最终认为直译能够更好地表达现场负责人对待安全严肃认真的态度,这样也能够让工人们更重视安全,事实证明确真如此。作为一名译员,必须意识到口译服务对象的性格千差万别,受教育程度也各不相同,要想与客户更好地共事,就必须了解不同客户的特点。

五、现场环境问题

为工程项目提供口译的译员常常会在口译现场遇到酷热难耐的天气,现场几乎无遮阴处,并且安全起见,所有进入现场的工作人员都需要穿戴厚重的防火服、安全帽、护目镜、迷你逃生器等安全装备,工程口译员需要克服现场恶劣的气候、环境等问题。除此之外,运转中的装置是极其危险的,除了可能会发生泄漏,可能还会有其他化学气体排出,有些设备还会散发冷、热蒸汽,而且装置运行时也会产生一定的噪声,这些都给口译工作带来困难和干扰。

第五节　工程口译员的必备素质

一、熟悉口译主题,了解相关知识

与其他项目不同的是,涉外机电项目的工作流程较为固定,其技术谈判和商务谈判的主要内容已基本包含在标书及附件里,所以口译前首先要做的就是了解标书文件的内容,并对关键内容进行深入研究。

机电项目的技术谈判主要集中在设备描述(description of equipment)、技术规格(technical specifications)、配置清单(configuration list)、服务范围(scope of services)四个方面。其中设备描述和技术规格是最难攻克的,因为其中涉及机电系统的工作原理。以某公司的中海油伊拉克双燃料柴油发电机组项目为例,在设备描述部分,业主的要求如下(节选):

The overall required power load at Intake Water Pump Station is approximately 2.2 MW. The Dual Fuel Diesel /Natural Gas Generator set supplied by VENDOR should be able to supply power to a 500 kW water pump for directly start. Vendor shall provide the diesel generator package with auxiliary equipment for the successful operation of the plant.

意思是:整个进水泵站所需的电力负荷大概为2.2兆瓦。供方提供的双燃料柴油/天然气发电机组应至少满足 500 千瓦水泵直接启动所需的功率。供方须为电站的成功运行提供发电机组和辅助设备。

同时附上其项目设计图供投标方参考。投标方的工程师在招标文件的设备描述部分根据招标方的需求进行如下设计(节选):

我司设计的电站设备包括:

(1)两台 40 尺集装箱式双燃料发电机组;

(2)两台 20 尺标准集装箱式卧式散热器;

(3)两台 20 尺集装箱式中、低压配电系统。

另附上电站的主接线方案。

作为初次接触机电领域的口译员,看到如此抽象的内容难免会摸不着头脑,无从下手,而这些内容又是技术谈判中着重讨论的部分,要想让双方的交流顺畅无阻,口译员必须了解其中的关联和逻辑。这时最有效、最快速的解决方法就是向工程师请教。请求工程师对图纸当中的各个部分的构成和工作原理进行细致的讲解。如:双燃料发电机组(Dual Fuel Generator Set)是指可以使用柴油和天然气的混合物作为燃料,也可以在缺乏天然气的情况下,只使用柴油运行的机组。而这里的机组不仅仅包括两台发电机,还包括启动系统、预热系统、供油系统和供气系统等,这些系统在图纸中一目了然,口译员可以很快掌握。40 尺集装箱选用全新 40 尺开顶高柜改制而成,其优点在于能满足海运、陆路运输及室外的快速就位。而之所以采用集装箱式卧式散热器是因为它可以保证机组在环境温度为 55℃的条件下可靠工作:散热器为全铜带式闭式内循环散热,两侧设有进风口及防护网,可适应沙漠地区的工作环境。中、低压配电系统能够实现对机组的送配电和对机组运行过程的监视控制。

译员对主题越了解,知道的信息越多,在翻译中的压力就会越小,口译质量才会越高。如涉及工程项目商务谈判,译员必须了解商务谈判的主要内容,主要包括价格清单(bid schedule of prices)、付款时间表(payment schedule)、合同进度及保证发货时间(contract schedule and guaranteed delivery date)以及必要保险(required insurance)、索赔(claims)等内容。此外,工程合同也是口译员在做现场口译前所要熟悉的。合同是双方协定的施工办法的依据,假设双方在施工过程中出现争议,合同条款就是最具说服力的工具。口译员若对合同有全面的了解,就更容易弄清楚双方的意思,从而进行准确而迅速的翻译。在执行现场口译任务之前,笔者都会先熟悉合同内容,以免出现不必要的差错。

二、掌握专业术语、缩略词及短语搭配

术语是指经过业界权威机构认定的专业名词,对学术界或其他领域的有效交流起到了重要作用。术语翻译在口译中的作用也是不可忽视的。世界主要口译行业组织,都对术语的准确性提出了要求,特别是国际会议口译员协会(AIIC),已经把术语翻译明确地列入口译质量评估标准中。在工程项目口译之前,我们有必要掌握相关的专业术语、缩略词。

以机电项目口译为例,译员需要掌握涉及承包工程的一些基本术语,如:EPC = Engineering-Procurement-Construction,设计—采购—建造总承包或称 EPC 总承包;PM = Project Management,项目管理;PMC = Project Management Contractor,项目管理承包商。

还比如:

Our company uses the special method for strengthening bearing surface, that is duplex hardening method.

翻译:我司运用了一种特殊的方法来强化轴承表面,即二次淬硬。

"duplex hardening"这个词语是新词,这一工业方法在世界上是领先的,所以词典中基本没有相应的汉语解释。后来多名专家根据其具体的工作运行方法(即先在轴承表面渗碳,再在其上渗氮,以增强轴承表面的硬度),并进一步研究与探讨之后,给它起名为"二次淬硬"。在工程口译中,译员不仅可以通过多种渠道了解并积累大量的技术词语,还能学习到很多相关的背景知识。

三、熟悉发言人的讲话特点

不同国家或不同地区的人所讲的英文在语音、语貌上都有很大的不同。例如:印度式英语发音的一个主要特点就是把标准英语中本应该咬舌送气的音"th"简化为"t",音"sh"简化为"s"。而且印度人发"t"的音,又接近"d"的音,"g"和"k"也是分不清的。例如:在一次发动机现场启动试验中,一位印度客户指着散热管道的一个接头说"Dell tem do change tas joint do a D-joint."。乍一听不知道他说的是什么,请他重复后才明白,他说的是"Tell them to change this joint to a T-joint",即"告诉他们把这个接头换成三通接头"。所以刚开始接触印度人的时候,适应他们的口音是比较困难的。澳大利亚口音常常省略辅音"h",直接发后面的元音。比如 how 经常发成 ow。澳大利亚人称朋友为 mate,而且喜欢夸张地发成 myte。澳大利亚人把 today 念成 to-die。在新加坡英语中,one、two、three 的发音是 one、true、tree。所以,在正式口译开始之前,口译员提前与与会者接触和交流是很有必要的。通过提前接触,口译员可对发言人的口音、语速和说话习惯有所了解,以便于准确地把握和理解源语内容,最大限度地减少接收和转化源语信息过程中所用的时间。

四、良好的心理素质

工程口译基本上在工作现场进行,有时需要面对数十人,因此对口译员的心理素质有严格的要求。口译员最好能够不卑不亢、不慌不忙、临危不惧、处变不惊。口译员刚开始也许会有些紧张,不过一旦专心投入工作中,紧张情绪就会缓解很多。通俗地说,口译员应当有副"厚脸皮",即使翻译错了,或者漏翻了,也不可表现出来,先让谈话继续,找到时机再补回来。

五、职业道德

口译工作直接涉外,因此口译员必须遵守职业道德,遵守外事和财经纪律,严格遵守保密协议,对自己的翻译内容负责。口译员不是外方的发言人,也不是中方的决策人,绝大部分时间充当的是双方的中介角色,因此任何时候都不能把自己当作主角,必须忠实地翻译双方的观点。

六、外交礼仪

工程口译员是连接中国与外国的桥梁,是外国了解中国的窗口,所以口译员在很大程度上代表着中国的形象。因此,工程口译员必须知晓并遵守基本的外交礼仪,相互尊重,不卑不亢,严格要求自己的语言和行为,为中外双方建立良好的关系做出重要贡献。

第六节　口译技能——多任务处理

一、训练目标

本单元以"影子跟读"及"小句填空"为练习方法,旨在训练学生在听辨中高度集中注意力,抗干扰,在不同任务间协调及有效分配注意力的能力。

二、技巧讲解

与笔译不同,口译中的信息接收转瞬即逝,信息处理、转换和产出几乎需要同时进行,没有斟酌和犹豫的余地。基于此,具有神经科学和认知科学背景的学者将口译看作一种高强度的认知负荷的活动,研究口译的信息处理过程,关注工作记忆管理和精力分配等主题,并提出了多个口译过程模型。其中,在口译教学界应用最广泛的是吉尔的认知负荷模型。他认为同声传译中,四个任务同时发生,即 simultaneous interpreting = listening + memory + production + coordination。

而在交替传译中,信息接收和产出分两个阶段,每个阶段涉及多个任务,例如:在接收阶段,consecutive interpreting = listening + memory + note-taking;而在产出阶段,consecutive interpreting = remembering + reading notes + production。

口译是典型的高强度多任务处理活动,这是由任务的即时性所决定的,听、记忆、做笔记等活动必须同时进行。同时,口译中的单个任务难度也很高,不属于自动完成范畴。以听为例,母语为中文的英语学习者通常需花费全部脑力来听辨,甚至有时听不懂。

那么,在高强度和难度并存的情况下,如何在口译中高质量地并线处理多个任务呢?与所有复杂技能的习得相似,强化训练是主要途径。通过反复练习,在任务和技巧间建立自然联系,促使过程中的各个子任务趋于自动化,以达到"不费力"的境界,即常说的"熟能生巧"。

本单元将介绍两种训练方法:影子跟读及小句填空。

训练方法一:影子跟读练习

影子跟读练习在传统上更多地运用于同声传译训练,对交替传译训练同样适用。在练习时,学习者用源语重复发言人的讲话,听、理解及跟读同时进行。跟读通常与发言人的讲话保持几个字词的距离,赢取理解的空间。在跟读完毕后,需简要回顾发言内容,或者回答教师根据发言内容提出的问题,以确保跟读时的信息处理没有停留在语音重复阶段。虽然内容回顾和记忆相关,但本单元训练不把记忆当作浮于意识上的主动任务,而是作为注意力高度集中和理解的自然结果,用以检验两者的效果。因此,影子跟读有三项子任务:

影子跟读 = 听 + 理解 + 跟读。

其中,"听"和"理解"为主要任务,"跟读"为次要任务,模拟交替传译中做笔记对听的影响,以及同声传译中同步转换对听辨的影响。由于重复跟读比做笔记和即时转换容易很多,因此能将任务难度控制在可处理范围。

影子跟读 = 听 + 理解 + 跟读 + 数字书写。

数字书写作为干扰项可以多种形式出现,如从 1 ~ 50 按顺序或倒序写数字,写具有特定规律的数字(基数或偶数等)。影子跟读越轻松的学习者越可采用复杂的数字书写方式。书写方式越复杂,占用精力越多,越能相应地激发学习者的脑力潜能,并进一步训练任务间的精力分配。

训练方法二:小句填空练习

小句填空练习要求学习者在听辨正常语速的发言时,填写以小句或长短句为单位的空白,并在材料播放完毕后,简要回顾材料内容。这一练习涉及三个子任务,其中填空为干扰项。

三、课后练习(音频:1.6)

小句填空练习:

This document stipulates the obligations and rights of the parties to the contract 1) _____ . In most cases, conditions of contract consist of two parts, that is, 2) _____ _____ . The employer would designate a standard form of contract like FIDIC Conditions of Contract as his general conditions of contract, 3) _____ . The name of the standard form and the version, however, will be stated in the instruction to bidders, or alternatively, 4) _____ _____ if he chooses to do so. Amendments made to the general conditions of contract according to specific project conditions constitute 5) _____ _____ .

第二章　工程项目启动

海外工程总承包项目合同一经签订,总承包企业应立即启动项目,任命专职项目经理,指导和协助项目经理组建项目部,派遣技术人员,设立驻外机构,调查工程所在国的社会、经济、法律、风俗习惯、宗教信仰等,了解工程所在国的基本物价水平、劳动力市场、机械租赁市场、材料多样性及采购的便利程度等。

第一节　背景知识

一、任命专职项目经理

项目经理是工程总承包项目的负责人,经过授权,代表工程总承包企业全面执行项目合同。项目经理除应满足资格、能力、知识、经验和道德等多方面的要求外,还必须是专职人员,因为工程总承包项目极其复杂,要求项目经理全身心投入。另外,对海外工程来说,项目经理最好具有较强的外语交流能力,以与业主及其他项目关系人顺利沟通,互相建立良好的个人关系。从前期工程实践看,目前的项目现场翻译绝大多数是刚毕业的大学生,既缺乏实践工作经验,又不具有工程的概念,难以满足日常交往、谈判工作中对专业理解力和准确表达能力的要求。

由于海外工程总承包项目的人力资源、分包商的选择、材料和设备供应等方面主要依托国内本部的支持,所以目前总承包企业项目经理的设置主要采用两种模式,一种是项目经理(常驻国外)+总承包企业部门(国内)支持的形式,另一种是现场经理(常驻国外)+项目经理(常驻国内)+总承包企业部门(国内)支持的形式。从现有实践看,前一种模式中,项目经理权力更集中,更有利于项目的高效实施,但它要求总承包企业相对成熟,有管理层和职能部门的强力支持。

二、组建项目部

工程总承包项目部的设立主要包括确定组织机构形式,明确职能和岗位设置,确定组成人员和职责权限,组织编制项目部的管理规定和考核、奖惩办法等内容。从当前海外工程实践看,工程总承包项目部的组织结构形式基本上为项目式结构(与常用的矩阵式结构有所不同),项目部的大部分人员为短期聘用,进企业后直接为特定项目部服务,部门的归属意识相对较弱,较少受到双重领导的干扰,有利于服从项目部的直接管理。

三、派遣技术人员

海外项目部成立后,应立即安排办理人员派遣的相关事宜,如护照的收集、访问签证和工作签证的办理、人员派出的最优(费用和时间)路线设计等。人员到场后,还应立即掌握暂住证及工作证的办理程序、往返签证以及签证延期的办理程序等。项目部应尽快建立起相关的管理办法,为即将到来的人员大批量调遣工作创造条件。

先期派出的人员必须至少包括项目经理(或现场经理)、翻译人员、土建工程师和采购工程师等,以满足现场各种决策、与外方沟通联络、施工条件调查、市场调查等各方面的需要。

众所周知,国外承包工程需要派遣有关技术人员和管理人员到国外去,这就涉及国际旅行的问题。因而,作为从事国际工程的有关人员就需要学习一些旅途用语和了解一些飞行常识,以便顺利地到达工程所在国。

一般说来,一个项目的人员往往分批而行,由于人多事杂,大家要相互照应,注意保管好自己的物品,尤其是护照和机票。登机前,要到有关航空公司的机场服务台凭机票和护照领取登机卡,并办理托运行李手续。注意:在行李标签上一定要填写清楚技术人员的姓名以及将行李托运到的地方。填写名字时,要注意英语中的 Given Name 是"名",而 Last Name/Family Name/Surname 则指的是"姓",以免取行李时遇到麻烦。

此外,航空公司对免费托运的行李重量有一定的限额,一般为七十磅(约为三十二公斤),超重的部分要缴纳托运费,而且费用很高。因此登机前要询问清楚,以免超重。另外,航空公司对每件行李的重量也有要求和限制。

在飞机上,用餐是免费的,饮料一般也是免费的,但是有些航班对供应的各种不含酒精的饮料不收费,对供应的啤酒却是收费的。在到达目的地前,机上服务人员会发给旅客两张表格,一张是旅客入境卡,另一张是海关申报单。在每个座位靠背后的袋子里一般都有旅客须知之类的小册子,里面有如何填写的说明及范例可供参考。下飞机后,旅客需先到机场移民局上交旅客入境卡,移民局官员随即检查旅客的护照、签证等个人入境信息,并询问相关问题。若没有疑问,便会让旅客入境。接下来旅客需要到领取行李处领取行李。最后再到海关上交海关申报单,办理入关手续。

如果工程技术人员需过境,且在过境国短暂停留几天后,还须换机飞往工程所在国,到机场候机大厅后,若所乘航班办理登机手续的服务台不好找,可向机场工作人员询问,他们将会很乐意地给予帮助。

四、设立驻外机构

一般情况下,项目部需要在工程所在国的国际机场所在地设立临时办事机构,负责人员接送、信息收集、市场调查等各种任务。该机构必须在首批人员到场时设立,在工程结束最后一批人员离场时才能撤销。办事机构的工作如下:

(一)调查工程所在国的社会、经济、法律、风俗习惯、宗教信仰等

除了在国内所能了解到的信息以外,首批派出人员到工程所在国后,还必须立即收集有关该国的社会、经济、法律、风俗习惯和宗教信仰等各方面信息,如出入境管理、劳动与用工、工程质量与安全、交通运输、宗教等的规定和限制。

(二)调查工程所在国的市场情况

调查了解工程所在国(地)的生活资料和常见材料的价格,并与国内市场价格进行比较;调查了解当地劳动力的供应能力、供应方式、供应价格以及其他有关规定;调查了解当地运输机械、土方机械和起重机械等的租赁市场情况,掌握租赁渠道(供应能力)、租赁价格等信息;调查了解当地专业市场上的工程常用材料的品种、规格、数量、质量及供货时间等方面的信息,与国内进行比较,以备不时之需。

第二节 相关词语与表达

序号	中文	英文
1	亚洲开发银行	Asian Development Bank
2	授权、委托	authorize(delegate)
3	工程量表	Bill of Quantities(BOQ)

续表

序号	中文	英文
4	安全监测仪器	safety monitoring instruments
5	土建工程	civil works
6	索赔	claim
7	商务经理	commercial manager
8	项目经理	project manager
9	施工管理	construction management
10	工程师代表	engineer's representative
11	业主	Employer (Client/Owner)
12	承包商	Contractor
13	外籍职员	Expatriate
14	土木工程师协会	Institution of Civil Engineers (ICE)
15	图纸	drawing
16	车间图	shop drawing
17	设计图	design drawing
18	竣工图	as-built drawing
19	蓝图	blue drawing
20	透明图	transparent drawing
21	施工图	construction drawing
22	电气工程	electric works
23	海外工程	overseas project
24	国内工程	domestic project
25	保险	insurance
26	劳务	labor
27	布置	layout
28	牵头公司、责任公司	leading company (sponsor)
29	责任、义务	liability (responsibility/obligation)
30	总包	lump sump
31	机械	machinery

续表

序号	中文	英文
32	人力资源	human resource
33	制造商	Manufacturer
34	材料	material
35	办法、措施	measure
36	测量、计量	measurement
37	备忘录	memorandum
38	进场	mobilization
39	反对	objection
40	支付	payment
41	设备	plant
42	资格预审	prequalification
43	采购	procurement
44	利润	profit
45	进度控制	progress control
46	项目经理	project manager
47	质量控制	quality control
48	申请、请求	request（application）
49	审查	review
50	风险	risk
51	截流	river closure
52	导流	river diversion
53	安全	safety
54	工地、现场	site
55	现场工程师	site engineer
56	规格、规范	specification
57	职员	staff
58	分包商	subcontractor
59	提交	submission

续表

序号	中文	英文
60	监督、监视	supervise
61	供货商	supplier
62	国际商会	International Chamber of Commerce
63	单价	unit price
64	变更	variation（change）
65	世界银行	World Bank
66	基本工资	basic wage
67	副食差价补贴	subsidy for price differential of non-staple food
68	施工津贴	construction subsidy
69	夜餐津贴	night meal subsidy
70	书报费	subsidy for readings
71	社会福利基金	social welfare fund
72	工会基金	labor union fund
73	劳保基金	labor protection fund
74	水、电、煤补贴	subsidies for water, electricity and coal
75	粮、煤、肉、菜差价补贴	subsidies for price difference of grain, coal, meat and vegetables
76	冬季供暖补贴	subsidy for heating in wintertime
77	其他补贴	other subsidies
78	劳保费用	expenditures for labor protection
79	住房津贴	subsidy for housing
80	管理费	overheads
81	进洞费	cost for entering into tunnels
82	进场、分派和解雇费用	cost for mobilization, assignment and dismissal
83	住房费	logging subsidy
84	标准工作时间外工作费	costs for work outside standard work time
85	闲置费用	cost of idleness
86	小工具及手工操作设备津贴	subsidy for small tools hand-operated facilities

续表

序号	中文	英文
87	中方营地保险费	costs for insurance of Chinese camps
88	非熟练工(1 级工)	unskilled labor (grade/class 1)
89	半熟练工(2 级工)	semiskilled labor (grade/class 2)
90	熟练工(3 级工)	skilled labor (grade/glass 3)
91	高级熟练工(4 级工)	highly-skilled labor (grade/class 4)
92	助理矿工	assistant miner
93	助理机修工	assistant mechanic
94	助理木匠	assistant carpenter
95	助理砼工	assistant concreter
96	助理厨师	assistant cook
97	吊车装卸工	assistant loader
98	卷扬机手	winchman
99	加油工	pump attendant
100	钻工	driller
101	炮工	blaster
102	矿工	miner
103	管工	plumber
104	装配工	rigger
105	喷砼工	shotcreter
106	装载机操作手	loader operator
107	水泵工	water pump man
108	自卸车司机	dumper driver
109	挖掘机操作手	excavator operator
110	推土机操作手	bulldozer operator
111	压路机操作手	compactor operator
112	搅拌运输卡车司机	transit mixer driver
113	卡车司机	truck driver
114	机械工	mechanic

续表

序号	中文	英文
115	电工	electrician
116	焊工	welder
117	钻头研磨员	drill sharpener
118	木工	carpenter
119	砼工	concreter
120	钢筋绑扎工	steel fixer
121	冲毛工	water blaster
122	试验人员	tester
123	机电安装工	hydromechanical installer
124	厨师	cook
125	砼泵工	concrete pump man
126	吊车操作手	crane operator
127	洞挖多臂钻操作手	tunnel jumbo operator
128	工长	foreman
129	会计师	accountant
130	助理工程师	assistant engineer
131	营地老板	camp boss
132	责任会计师	chief accountant
133	主厨	chief cook
134	责任绘图员	chief draftman
135	责任购物员	chief purchaser
136	责任计量师	chief quantity surveyor
137	计算机设备人员	computer equipment personnel
138	日工资费率	daily rate
139	劳工部	department of labor
140	医生	doctor
141	野外工程师	field engineer
142	1班	first shift

续表

序号	中文	英文
143	整日制工作	full employment
144	会计长	general pay master
145	节假日加班	holiday overtime
146	小时费率	hourly rate
147	女仆	house maid
148	口译人员	interpreter
149	初级会计师	junior accountant
150	初级工程师	junior engineer
151	初级计量师	junior quantity surveyor
152	初级测量师	junior surveyor
153	试验工程师	laboratory engineer
154	劳工市场	labor market
155	劳工人员	labor officer
156	当地工人	local labor
157	当地人员	local personnel
158	劳工部	ministry of labor
159	加班小时	overtime
160	出纳员、发款员	pay master
161	发薪日	pay-day
162	工资核查人员	payroll checker
163	工资支付人员	payroll officer
164	工资单	payroll（paysheet）
165	薪资核算单	pay-slip
166	计件工作	piecework
167	公关人员	public relationship officer
168	采购人员	purchaser
169	安全人员	safety officer
170	季节工作	seasonal work
171	2班	second shift
172	秘书	secretary

续表

序号	中文	英文
173	保安人员	security assistant
174	高级工程师	senior engineer
175	高级计量师	senior quantity surveyor
176	高级测量师	senior surveyor
177	换班工作	shift work
178	仓库会计	store accountant
179	3班	third shift
180	计时工作	time-work
181	按小时付报酬	to be paid by hour
182	到达现场月份	arrival month at site
183	正常使用条件	average service conditions
184	退货运费	back freight
185	提货单	bill of lading
186	铁路提货单	railway bill of lading
187	移民安置	resettlement of immigrants
188	运费已付	carriage paid
189	运输	carriage（traffic, transportation）
190	运货费	cartage
191	租用条件	conditions of hire
192	托运	consign
193	建筑设备公司	construction equipment company
194	施工机械	construction plant
195	运输合同	construction of carriage
196	运送	convey
197	设备资金费用	cost of facilities capital（CFC）
198	油料费	cost of fuel
199	润滑油费	cost of lube
200	大修费用	cost of overhaul
201	大修配件费	cost of overhaul parts

第三节　对话口译

Dialogue 1:公司推介（音频:2.3.1）

Word Tips

江西省水利水电建设集团有限公司	Jiangxi Water and Hydropower Construction Group Co. Ltd.
市场营销工程师	marketing engineer
名片	business card
房屋建筑工程	building works
土木工程	civil works
管道工程	pipeline engineering
国际承包人	international contractor
年营业额	annual turnover

【江西省水利水电建设集团有限公司的项目经理张鹏(以下简称 Z)向客户代表 James(以下简称 J)介绍自己的公司。】

Z:您好!

J:How do you do?

Z:我是张鹏,江西省水利水电建设集团有限公司的项目经理。这是我的名片。

J:Thanks. Here is mine.

Z:请允许我简要介绍一下我们公司。

J:Sure.

Z:这是我们的宣传册。

J:The pictures are beautiful.

Z:谢谢您! 我们公司承建过许多重要工程,宣传册中只显示了一部分。对了,我们公司成立于1956 年。

J:So your company has been in operation for many years.

Z:是的,我们在过去的68 年里创造了辉煌的业绩。

J:Is your group focusing only on the building works?

Z：尽管我们以承包水电站工程著称，我们也涉足其他土木工程，例如公路、桥梁和管道工程等。现在我们正在寻找机会承接环境保护工程。

J：Do you have any experience working abroad or in connection with international contractors?

Z：有。我们集团在东非、西非和东南亚承接过很多施工项目。我们同来自法国、德国和日本等国的承包人合作也非常顺利。

J：What about your registered capital?

Z：大约 3 亿元人民币。

J：Great! I hope we will have the opportunity to cooperate.

Z：期待能尽快与您和贵公司合作。

Dialogue 2：全球工程承包市场（音频：2.3.2）

Word Tips

Public Works Bureau	工务局
global construction market	全球建筑市场
international engineering & contracting market	国际工程承包市场
financial crisis	金融危机
sovereign debts crisis	主权债务危机
economic momentum	经济势头
the Belt and Road Initiative	"一带一路"倡议
capital investment	基本建设投资
F + EPC（Finance + Engineering-Procurement-Construction）	融资加设计—采购—施工模式
PPP（Public-Private Partnership）	政府和社会资本合作模式
BOT（Build-Operation-Transfer）	建造—运营—移交模式
financial guarantee	融资担保
AIIB（Asian Infrastructure Investment Bank）	亚洲基础设施投资银行
the Export-Import Bank of China	中国进出口银行
China Development Bank	国家开发银行
Industrial and Commercial Bank of China	中国工商银行

Silk Road Fund	丝路基金
China-Africa Development Fund	中非发展基金
China-Africa Fund for Industrial Cooperation	中非产能合作基金

【场景:江西省水利水电建设集团有限公司的项目经理张鹏拜访老朋友——当地工务局领导安德森先生,并就全球工程市场情况相互交流。】

Mr. Anderson: Mr. Zhang, my old friend. Long time no see. How are you?

张鹏:安德森先生! 我很好啊,你呢?

Mr. Anderson: Very well. I heard that you have been promoted to vice president of your group. Congratulations! I am not surprised given your outstanding achievement and performance in international construction.

张鹏:谢谢您的赞美!

Mr. Anderson: China has been developing rapidly in past decades, and the share of Chinese contractors in the global market is also growing fast. It is amazing!

张鹏:很难相信自从我们上次见面已经过去十年了。中国加入世界贸易组织也有 20 多年的时间了。自那时起,中国承包商在国际工程市场上快速成长。我公司在中国政府"走出去"战略的指引下也取得了巨大的成就。不过坦率地说,我愿意把我们公司面临的一些瓶颈和问题也告诉你。借此机会,我能否有幸聆听你的远见卓识,并了解你对未来国际工程承包市场前景的看法呢?

Mr. Anderson: Well, yes, I have noticed some changes in the international engineering & contracting market, which I would be very happy to discuss with you. But why don't you describe your situation first?

张鹏:好极了! 我们公司的业务集中在亚洲、非洲以及拉丁美洲的欠发达国家。但受到 2008 年金融危机和 2012 年欧洲债务危机的拖累,这些国家的经济一落千丈、一蹶不振,严重影响到国际贸易和国际工程市场。由于原油和矿产价格的下跌,严重依赖资源出口换取外汇的欠发达国家受到沉重的打击,他们的财政收入乃至建设预算都大幅下降。很遗憾我们有几个项目的付款都遭到严重拖延。最近我们注意到,新项目一般都要求承包商具有一定的投资和融资能力。

Mr. Anderson: I also noticed this trend. It seems these difficulties are not only encountered by Chinese contractors, but also by other contractors all over the world.

I believe that among Asia, Africa, Central and Eastern Europe, and Latin America, the Asian economy is much better than the other regions comparatively.

张鹏:哦,对我们来说,中东地区也是我们的重要客户,但中东局势动荡,许多国家的形势极其紧张和不可预测。例如,叙利亚战争持续了多年,再加上巴以冲突并没有得到解决。

Mr. Anderson: Yes, but if you just consider the Central and Southeast Asian countries, and India, they seem to be resuming their economic momentum and showing great development potential. I believe their governments' ability to fund projects and make prompt payments.

张鹏:太好了,我们俩的观点一致。这些国家与中国毗邻,一直是我们的主要市场。但你知道,那里的竞争变得更加激烈了。

Mr. Anderson: I agree with you. In my opinion, Africa is the most potential continent with its vast territory and rich resources. Most of the African countries are still at the primary stage of economic development. So they have a strong focus on basic infrastructure development and how to enhance the opportunity for their people to have a good livelihood.

张鹏:我们公司在非洲也有很多项目,包括一些由中国政府援建的项目,以及更多由中国的银行提供融资的项目。

Mr. Anderson: The Latin America market is growing fast too, but you need to pay close attention to the market risk.

张鹏:你是对的! 由于大宗商品价格低迷,许多依赖自然资源的发展中国家面临着艰难的境地。由于基本建设投资和融资担保能力的大幅下降,采用 F+EPC 模式的项目越来越少,而采用 PPP、BOT 模式的建设项目则越来越多。在这一新的市场环境下,我公司正努力进行业务转型,以适应当前的形势。

Mr. Anderson: The Chinese government was far-sighted in launching the Belt and Road Initiative. This has played a strong role in stimulating the economies of many countries and created many opportunities in the international contracting and engineering market. It also led the establishment of the Asian Infrastructure Investment Bank, to support infrastructure construction in developing countries. I truly admire this.

张鹏:中国政府提出的这一倡议对促进国际工程承包行业的发展具有重要作用。我公司与中国进出口银行、国家开发银行等政策性银行及中国工商银行、中国银行等国有商业银行都有密切的合作,协助外国政府融资,推动相关项目的落地。同时,我们还与丝路基金、中非发展基金、中非产能合作基金等各类基金有密切联系,来实施投资项目、PPP 项目和 BOT 项目。

Mr. Anderson: That's very inspiring to hear. As a matter of fact, the developed countries are also facing the problem of lack of funds for infrastructure reconstruction. I hope you can come to my country to investigate and conduct business there in PPP model. I can introduce you some of my good friends who might be very helpful to your business start-up.

张鹏:非常感谢你,安德森先生。我很乐意在贵国与您合作!

第四节 篇章口译

Passage 1:英译中(音频:2.4.1)

Word Tips

Construction Organization Plan	施工组织计划
deploy	配置
accredit	委派
site inspection	现场检查
terrain	地形
access routes	通道
stakeholders	利益相关者
structured approach	结构化方式

【场景:项目经理张鹏向甲方介绍施工计划。】

Good morning, everyone! First of all, please allow me to introduce myself. My name is Zhang Peng, the contraction manager of this project and I'll take charge of everything related to construction. Now, I'd like to tell you briefly about our construction plan. We will go all out to make pre-construction preparations and the examinations of all the schemes for the project execution.

Construction Organization Plan (COP) is a detailed and comprehensive document that outlines the planning and execution of a construction project. The main purpose of COP is to ensure that the construction work is carried out smoothly, safely, and efficiently with minimal disruption to the surrounding environment. A well-prepared COP can help to reduce risks, control costs, and deliver the project on time.

Step 1: Site inspection and evaluation

The first step in preparing a COP is to conduct a thorough site inspection and evaluation. The purpose of this inspection is to identify any potential hazards or obstacles that may affect the construction work. The evaluation should cover factors such as soil conditions, existing structures, terrain, access routes, drainage systems, and utilities.

Step 2: Identify the project team

Once the site evaluation is complete, the next step is to identify the project team. This team will be responsible for developing the COP and ensuring that it is implemented correctly. The project team should be composed of individuals with relevant expertise and experience. The team should include a project manager, construction manager, engineers, safety officers, and other key personnel.

Step 3: Develop the COP

The next step is to develop the COP. The COP should include detailed information on the project scope, schedule, budget, construction methods, quality control, safety procedures, and risk management. The COP should also outline the roles and responsibilities of the project team and any subcontractors involved in the construction work.

Step 4: Consult with stakeholders

It is important to consult with relevant stakeholders when developing the COP. This includes local authorities, regulators, and the community. Consultation with stakeholders can help to identify any issues or impacts related to the construction work. This information can then be incorporated into the COP to ensure that the project is carried out in a socially responsible and sustainable manner.

Step 5: Review and adjust the COP

Once the COP is complete, it should be reviewed and adjusted as necessary. The objective of this review is to ensure that the COP is comprehensive and effective. The review should consider factors such as changes in the project scope, site conditions, or regulations. Adjustments to the COP may be required to ensure that the project is carried out safely, efficiently, and meets the required quality standards.

In conclusion, the Construction Organization Plan is a critical component of any construction project. It enables the project team to plan and execute the construction work in a safe, efficient, and sustainable manner. The COP should be developed using a structured approach that involves a thorough site evaluation, the identification of key personnel, consultation with stakeholders, and regular review and adjustment.

Passage 2:中译英（音频:2.4.2）

Word Tips

civil works	土建工程
coal fired power station	燃煤发电站
substructure	基础结构
superstructure	地上结构
National Prize for Progress in Science and Technology	国家科技进步奖
pre-qualification document	资格预审书
subsidiary	附属机构

【土建承包人代表张鹏在招标会上介绍其所在公司的概况。】

早上好！我是张鹏,在江西省水利水电建设集团有限公司工作,我是土建承包人代表。首先,感谢你们给我公司提供了这次机会。接下来,我想向你们介绍一下我公司。我公司是全球国际承包商250强之一、江西省"走出去"十大领军企业之一,在水电站建设、公路工程、建筑工程、机电工程施工方面有非常丰富的经验。

在国际市场上,我们公司作为参与国际业务较早、国际化程度较高的施工企业之一,不断加快"走出去"步伐,大力推进"大海外战略"。集团在肯尼亚、埃塞俄比亚、莫桑比克、乌干达、贝宁等十余个国家设立了分公司和办事机构，积极开展对外投资和工程建设项目，对 PPP、EPC 等新模式具有较强的综合运营管理能力。

公司自 1993 年开始涉足海外工程，参与国际市场竞争，承建了埃塞俄比亚哈利河、厄尔巴耶、格力拉水利工程，埃塞俄比亚 WOLKITE 大坝沥青心墙工程项目、隧洞项目，埃塞俄比亚 Ribb 灌溉项目，肯尼亚双河商场项目，肯尼亚 3 个小水电站（EPC），肯尼亚内罗毕电网升级改造工程，肯尼亚维多利亚银行项目，肯尼亚 KENHA 办公楼项目，肯尼亚阿尔玛公寓楼项目，越南南化水电站，蒙古国公路工程，津巴布韦宾督拉大坝工程，突尼斯两座水坝项目，汤加王国医疗保健中心建设项目，阿富汗帕尔旺水利项目，印度尼西亚帕卡特水电站项目，莫桑比克 Corumana 大坝项目，乌干达国防医院项目，卢旺达鲁苏摩水电站项目，卢旺达 MIG 商业综合体项目等近百项工程。

近年来，随着"一带一路"倡议的提出，公司通过"大海外战略"的实施，逐渐实现了海外市场开拓和海外建设参与方式的多样化，承接业务和完成产值稳步上升。在市场开发方面，由项目带动市场的模式转变为结合"一带一路"倡议和各类政策主动布局海外市场，通过立足东非辐射整个非洲。与此同时，公司加大丝绸之路经济带沿线国家——巴基斯坦、缅甸、乌兹别克斯坦、哈萨克斯坦、格鲁吉亚等的市场开拓。此外，公司还以印度尼西亚为基点站稳东南亚市场，通过对外投资的方式实现业务增长多元化。

我们将在资格预审书后附上一些关于我公司详细信息的小册子。您将通过这些小册子详细了解我公司总部、子公司及附属机构的组织构成，以及我公司拥有的施工机械设备。

第五节　口译技能——主旨听辨

一、训练目标

本单元聚焦口译听辨阶段,旨在训练学生在多任务处理情境下对注意力的优先分配能力,要求能够准确识别发言人的意图,提取核心信息。

二、原理及技巧讲解

口译中的"听"是积极听辨,即译员不仅要接收输入的语音信息,而且要主动分析处理信息。具体来说,听辨活动涵盖的子任务包括但不限于声音识别、音素识别、语义提取、意义理解、信息处理等,其最终目的是使译员的翻译能够接近100%的原文信息。口译任务的即时性和篇幅过长无疑加大了听辨活动的难度。在一些口译场景中,译员的翻译会与讲话者的原意背道而驰,甚至会出现一些违背常识的话语。除去背景知识造成的困难外,其中很大原因归咎于译员没有很好地理解发言目的。初学者更容易犯这种错误,主要体现在只能够捕捉到一些信息碎片,翻译时无法把它们组织成逻辑清晰的话语。要避免这种情况出现,听辨训练就要先从宏观角度入手,把握话语的特征,准确识别发言人的意图。

从语言活动本身来看,发言基本上都带有一定的目的性或方向性,这一观点已被语言学家证实。文本语言学家 Beugrande 和 Dressler 认为,意图是任何文本(口头或书面)的一个固有属性;他们将其定义为文本生产者为实现计划中的某个目标表现出的态度及意图。在需要口译辅助的发言场合,发言的意图尤为明显。例如:产品发布会上,发言人会通过描述产品的各方面特征来达到吸引潜在客户的目的;辩论活动中,正反双方需要用大量论据支撑自己的立场;大型国际会议的圆桌讨论环节都会设立一定的主题,发言人需表明自己的立场,并展开讨论。

因此,我们可以把发言人的意图定义为发言人希望传递的主旨信息或达到的交流目的。发言人的意图可能会在话语中明晰化,也可能会隐藏在话语之中。口译员的首要任务就是识别其意图,并用以引导后续的听辨过程。识别出发言人的意图后,译员即使在后续的听辨过程中迷失了方向,也可以重新回到正常的轨道。

鉴于口译听辨也是多任务处理过程,本单元的训练要求学习者把识别发言人的意图放在其他子任务之上,提取发言的核心信息。

那么,如何准确识别发言人的意图?在此提出两个技巧:

(1)适当了解发言规范。所谓发言规范,指的是发言人在语言组织过程中会适当利用一些标记语。以英语发言为例,发言人经常在一开始讲话就提到要讲什么内容,并发出明显信号,如"What I would like argue is..." "What I want to share with you today is..." "The main purpose of my talk is..."。要特别注意的是,这些标记通常会出现在语篇或语段开头,语篇或语段展开遵循"总—分"规则。如果发言中存在这些明显信号,译员尽早将其捕捉到就能尽快识别出发言人的意图。

(2)利用语境信息并激活已有背景知识来预测。所谓语境信息,即讲话者、听众、发言场合、发言主题等交际信息,而已有背景知识指的是译员自身对交际信息的了解。每次听辨前都可以问自己一些问题:何人在何时何地对何人说话?活动主题和讨论话题是什么?为什么发言人要说这些话?掌握并有效调用语境和背景知识是口译的重要基础之一,也是译员能力的重要组成部分。

三、课后练习(音频:2.5)

概括下文主旨:

Needless to say, both the contractor and the employer are profit driven. Generally, the technical standards or criteria stipulated in contract documents would be the minimum requirement for the contractor in terms of quality control. All the contractors in the world tend to satisfy the requirements with minimum input. On the contrary, the employer would expect the best quality product despite the minimum requirement stated in the contract, since he has made an assumption that he would pay the highest price under a lump sum contract. The tolerance for the project quality has always been problematic and conflicts over such issues are inevitable.

第六节　课后练习

（音频:2.6）

Ⅰ. Technical Terms Interpretation

1. 国际承包人

2. 年营业额

3. 国际工程承包市场

4. 主权债务危机

5. 图纸

6. 蓝图

7. "一带一路"倡议

8. 配置

9. 现场检查

10. 工务局

11. F + EPC（Finance + Engineering-Procurement-Construction）

12. project director

13. economic momentum

14. terrain

15. the Silk Road Fund

16. civil works

17. pipeline engineering

18. Construction Organization Plan

19. access routes

20. structured approach

Ⅱ. Sentences Interpretation

1. 您先好好休息一下,明天我们再去项目现场。

2. 尽管我们以承包水电站著称,但是我们也涉足其他土木工程,例如公路、桥梁和管道工程等。

3. 现在我们正在寻找机会承接环境保护工程。

4. 我公司是中国最大的建筑公司之一,在燃煤发电站、核电站和水电站的建设方面有非常丰富的经验。

5. 作为总承包人,我公司去年在四川建了一座 2×600 MW 的燃煤发电站。

Ⅲ. Dialogue Interpretation

【中方项目总监的助手李想(以下简称 L)在机场迎接外方专家 Neil Williams(以下简称 N)。】

L:请问您是威廉姆斯先生吗?

N:Yes. That's right. I'm Neil Williams. Just call me Neil.

L:您好! 尼尔,我叫李想,是项目总监的助手。

N:How do you do? Glad to meet you, Li Xiang.

L:您知道的。项目总监张总让我接您去酒店。

N:Thank you!

L:您旅途顺利吗?

N:Not very good. I got airsick.

L:您现在觉得好点了吗?

N:Yes, but I feel a little tired, because it was a very long trip.

L:那我们走吧。我们这就去酒店办理入住手续。您先好好休息一下,明天我们再去项目现场。

N:OK. Thank you so much!

L:请在这里等我一下,我要去停车场取车。

N:OK. No problem.

Ⅳ. Passage Interpretation

The material are the basic elements of any building. Building materials may be classified into three groups, according to the purposes they are used for. Structural materials are those that hold the building up, keep of rigid, form its outer covering of walls and roof, and divide its interior into rooms. In the second group are materials for the equipment inside the building, such as the plumbing, heating and lighting systems. Finally, there are materials that are used to protect or decorate other materials.

第三章　工程项目管理之招标投标管理

　　国际工程通常是指一项允许有外国公司来承包建造的工程项目,即面向国际进行招标承包建设的工程。招标和投标作为比较成熟的、高级的规范化贸易方式,是目前国际工程领域普遍应用的、有组织的市场交易行为。了解国际招标、投标运作方式和掌握国际招标、投标技术对于从事国际工程承包的公司参与国际工程领域的竞争具有重要的意义。

　　国际工程承包是一项综合性的国际经济技术合作方式。它是指从事国际工程承包的公司或联合体通过招标与投标的方式,与业主签订承包合同,取得某项工程的实施权利,并按合同规定,完成整个工程项目的合作方式。因为工程资金金额较大,技术性较强,对工期、质量亦有较高要求,所以工程的发包人一般采用国际竞争性招投标来选择合适的承包人,以保证中标者在经济上和技术上有实力。

　　通过国际承包工程,可以实现技术、劳务、设备及商品等多方面的出口,不仅能多创外汇,而且具有一定的政治影响。中国对外承包工程业务发展很快,在"守约、保质、薄利、重义"八字方针的指导下进行。

　　随着经济合作和技术交流规模的不断扩大,我国建筑施工综合实力的增强,我国涉外单位和建筑施工队伍走出国门并参与国际工程承包市场竞争势在必行。要想在激烈的市场竞争中脱颖而出,做出既合理又有竞争力的投标报价是国际工程承包商参与竞标并中标的关键。

　　国际工程报价投标工作涉及面很广。它要求投标报价的人员具有广博的知识和丰富的经验,其中包括专业知识,如设计、施工组织与技术、材料设备等。同时,投标报价人员还必须具备商务方面的知识,如法律、金融、税务、保险,并熟知当地的政治、经济状况等。另外,投标报价人员还必须熟悉国际工程施工和投标报价的规范和操作程序。只有这样,投标报价人员才能在激烈的国际工程承包市场竞争中,为自己所在的承包公司赢得中标的机会。

第一节 背景知识

一、招标的承包合同种类

（一）按承担责任划分

1. 分项工程承包合同。发包人将总的工程项目分为若干部分，发包人分别与若干承包人签订合同，由他们分别承包一部分项目，每个承包人只对自己承包的项目负责，整个工程项目的协调工作由发包人负责。

2. "交钥匙工程"承包合同。交钥匙工程指跨国公司为东道国建造工厂或其他工程项目，一旦设计与建造工程完成，包括设备安装、试车及初步顺利运转后，即可将该工厂或项目所有权和管理权的"钥匙"依合同完整地交给对方，由对方开始经营。因而，交钥匙工程可以看成是一种特殊形式的管理合同。服务单位需要按市场经济规律，本着互惠互利、相互促进及相互支持的原则完成交钥匙工程。要承担交钥匙工程，服务单位没有一定经济实力是不行的。

3. "半交钥匙工程"承包合同。承包人负责项目从勘察一直到竣工后试车正常运转，确保产品符合合同规定标准，即可将项目移交给发包人。它与"交钥匙工程"承包的主要区别是不需要负责一段时间的正式生产。

4. "产品到手工程"承包合同。承包人不仅负责项目从勘察一直到正式生产，还必须在正常生产后的一定时间（一般为两三年）内进行技术指导和培训、设备维修等，确保产品符合合同规定标准。

（二）按计价方式划分

1. 固定价格合同，或称总包价格合同。固定价格合同是指在约定的风险范围内价款不再调整的合同。双方需在专用条款内约定合同价款包含的风险范围、风险费用的计算方法以及承包风险范围以外的合同价款调整方法。

2. "成本加费用"合同。"成本加费用"合同是指承包人垫付项目所需费用，并将实际支出费用向发包方报销，项目完成后，由发包人向承包人支付约定的报酬。

二、招标程序

招标过程分招标准备阶段、招标阶段和决标成交阶段。

（一）准备阶段的主要工作

选择招标方式、办理招标备案和编制有关招标文件。

(二)招标阶段的主要工作

发布招标广告、投标者资格预审和编制发售招标文件。

(三)决标成交阶段的主要工作

开标、评标和定标,定标后要颁发中标函、授予合同和签署合同协议。

三、招标文件

招标文件是指业主颁发给投标者的文件,一般包括合同条件、规范、图纸、工程量表、投标书格式以及投标者须知。在招标阶段编制和发售招标文件时,会涉及大量文案工作和沟通工作,包括招标文件的编制、发售与修改,邀请函的印发,投标者须知的告知,合同条件包括通用条件和专用条件的制定,以及投标书格式、图纸、技术规格书、工程量表、附加资料表、资料数据的选定。另外,要确定回函信封,即递送投标书的信封——为了确保准确无误地送达,信封上面印有业主的地址。最后,组织现场考察、招标答疑(如有疑问可向业主提出质询)、接收投标文件。

四、国际工程招投标

国际工程的招投标程序较国内普通工程招投标程序更为复杂。

(一)国际工程招标的程序包括确定招标方式、确定招标程序(按国际招标基本程序进行)、做好招标前准备工作、开标、评标、决标和授标。另外,国际工程投标的程序包括投标决策、组建投标班子、做好投标前准备工作、计算标价、确定投标策略、编制标书、做好投标过程中的细节工作以及签订合同等步骤。

(二)国际工程招标与投标的类型可分为国际竞争性招标、国际有限招标、两阶段招标和议标四类。国际竞争性招标是指在国际范围内,对一切有能力、有资格的承包人一视同仁,凡有兴趣的均可报名参加投标,通过公平、公开、无限竞争,择优选定中标人。采用国际竞争性招标可以最大限度地挑起竞争,形成买方市场,使招标人有最充分的挑选余地,获得最有利的成交条件。因此,国际竞争性招标是目前世界上使用最多的成交方式。

第二节 相关词语与表达

序号	中文	英文
1	资格预审	prequalification
2	招标	tender
3	建设	construction
4	持续的时间	duration
5	清偿,结算,清盘	liquidate
6	滞留,保持,保留	retention
7	保养,维修,维护	maintenance
8	图纸,图示	drawing
9	正式批准,正式生效	ratify
10	招标文件	the tender document
11	建设工程合同条件	the contract conditions
12	误期损害赔偿费	liquidated damages
13	招标图纸	tendering drawings
14	招标须知	instructions to tender
15	滞留金百分比	percentage of retention
16	保修期	maintenance period
17	中标	win the bidding
18	毁坏,拆毁	demolition
19	垦荒	reclamation
20	扩建,扩展	extension
21	开工日期	commencement date
22	连署,副署	counter-sign
23	适用的	applicable
24	规定,条款,条件	provision
25	自……生效	valid from
26	遵守,服从,符合	comply
27	减轻,缓和	mitigation
28	达到……满意程度	to the satisfaction of

续表

序号	中文	英文
29	环保机构	environment agency
30	初步进度计划	preliminary
31	承包商,立契约者	contractor
32	地方当局	authority
33	履约担保	performance bond
34	接受证书,移交证书	taking-over certificate
35	营业执照	business license
36	施工方案	method statement
37	先后次序	sequence
38	方法论	methodology
39	条件,限制性条款	proviso
40	如果,一旦	in the event that
41	偿还,补偿	reimburse
42	收信人,由……处理	for the attention of
43	拆除与填筑工程	demolition and reclamation works
44	以……为条件	subject to
45	合同价	contract sum
46	固定包干价	fixed lump sum
47	环境影响评估	environment impact assessment(EIA)
48	无异议证书	No-objection Certificate (NOC)
49	承包商一切险保险单	contractor's all risks policy
50	有限责任公司	Limited Liability Company (L. L. C)
51	专攻	specialize
52	居住的,住宅区的	residential
53	原始的	primeval
54	预制	prefabricate
55	拆开,拆卸	dismantle
56	报价	quotation
57	最后定下来	finalize
58	建成可交付使用	turn-key

续表

序号	中文	英文
59	附录	the attached appendix
60	编号为……的补遗	the addenda（numbered as...）
61	修补,补救	remedy
62	遵从,符合,一致	conformity
63	符合,按照	in conformity
64	通用合同条款	general conditions of contract
65	特别合同条款	special conditions of contract
66	预审书	prequalification document
67	土建工程	civil works
68	FIDIC 合同条件	FIDIC conditions
69	单价合同	unit price contract
70	工程范围	scope of works
71	施工进度计划表	construction schedule
72	技术规范	technical specifications
73	质量保证	quality assurance
74	投标专用	for tender use only
75	协议书	letter of agreement
76	施工方案	construction method statement
77	原本	original copy
78	书面委托书	written authorization
79	改动	alteration
80	行间书写	interlineation
81	保证金	advance payment
82	授予合同	award of contract
83	投标押金	bid bond
84	投标费用	bid cost
85	投标商	bidder
86	招标截止日期	date of the closing of tender
87	投标日期	bid due date
88	投标评估	evaluation of bids

续表

序号	中文	英文
89	招标	bid invitation
90	核标	examination of bid
91	投标价	bid price
92	合同金额	contract amount
93	投标书澄清	clarification of bid
94	招标书有效期延长	extension of validity of bid
95	投标书	form of tender
96	授权书	power of attorney
97	工程量计价清单	priced bill of quantities
98	评标标准	evaluation criteria
99	标准招标文件	standard bidding document
100	投标书的提交	submission of bid
101	标的物	subject matter
102	换标	substitution of bid
103	决标	tender decision
104	议标	tender discussion
105	招标方	tenderee
106	投标方	tenderer
107	投标书有效期	validity of bid
108	撤标	withdrawal of bid
109	做标	work out tender document
110	有竞争力的投标	competitive bid
111	不可预见费	contingency fund
112	中标合同	contract awarded
113	贷款信用保函	credit guarantee
114	纯损失	dead loss
115	不接受报价	decline an offer
116	最后期限	deadline
117	出口许可证	export license
118	财务能力	financial capacity

续表

序号	中文	英文
119	财务状况	financial position
120	备好投标文件	finishing bid
121	担保书	guaranty bond
122	保函	guarantee letter
123	担保人	guarantor
124	高级谈判	high-level talks
125	初步价格	initial price
126	保险标的	insurance subject
127	保险费	insurance premium
128	发行招标文件	issue bidding document
129	出具保函	issue guarantee
130	贷款协议	loan agreement
131	最低价投标人	lowest bidder
132	最低报价	lowest bidder（lowest tender）
133	标准格式	master format
134	投标文件的修改	modification of bids/tender
135	公证员	notary
136	公证书	notarial certificate
137	谈判	negotiated bidding（negotiating tender）
138	报价	offer in a bid
139	报价人	offerer
140	公开询价	open inquiry of the offer
141	参加投标	participate in a tender
142	履约保函	performance security
143	出口港	port of exit
144	纯利	pure interest
145	公平价格	reasonable price
146	重新投标	re-bid（retender）
147	投标审查报告	report on tenders
148	撤销保函	restitution of the guarantee

续表

序号	中文	英文
149	投标的密封和标志	sealing and marking of bid
150	选择性招标	selective bidding
151	单一投标(单项招标)	single tender
152	保函格式	specimen of letter of guarantee
153	财务状况报表	statement of financial position
154	中标者	successful bidder
155	大写金额	sum in words
156	投标委员会	tender committee
157	投标结果	tender result
158	报价截止日	the closing date of the offer
159	不平衡报价	unbalanced bid
160	不确定因素	uncertain factor
161	不公平竞争	unfair competition
162	不可预见因素	unpredictable element
163	招标代理	tender agent
164	招标资料表	tendering data sheet
165	直接签订合同	direct contracting
166	公开招标	open tendering
167	投标货币	tender currency
168	采购主体	procuring entity
169	开标记录	bid opening record
170	废标	fled bidding
171	履约期限	deadline for performance
172	保密协议	confidentiality agreement
173	备选方案投标	alternative bid
174	不可抗力	force majeure
175	采购代理	procurement agent
176	采购决定	procurement decision
177	到岸价	cost, insurance, and freight (CIF)
178	工厂交货价	Ex Works (EXW)

第三节 对话口译

Dialogue 1（音频:3.3.1）

Word Tips

an invitation for bid	投标邀请书
the bid opening	开标

Zhang Tian: Hi, Mr. Chen. I'm Zhang Tian and I was recruited a month ago by the HR department.

陈亮:您好,张先生。很高兴见到你。

Zhang Tian: Glad to see you, too. I know you're the project manager of the West African region.

陈亮:是的,确实如此。

Zhang Tian: As a newcomer, I'm eager to know something about an invitation for bid (IFB) because I want to make myself available for my post as soon as possible. Can you spare me a few minutes?

陈亮:我的荣幸。你想知道什么?

Zhang Tian: I was told that there are many notices in the element of general information and requirements of an IFB. What're they?

陈亮:嗯,通常,它们是:1)投标价格必须在开标后的指定天数内保持稳定;2)关于投标意外截止;3)关于您所在司法管辖区取消或拒绝投标的权利。

Zhang Tian: I see. How long should the bid prices be remained after the bid opening?

陈亮:这要看情况。例如,我们下周一发布 IFB,通知投标人,投标价格应在35 天内保持稳定。

Zhang Tian: OK, thanks for your explanation. I think someone is calling you and I would come again next time.

陈亮:不客气。您可以在周二、周三和周四与我联系。

Zhang Tian: Many thanks, bye.

陈亮:再见。

Dialogue 2（音频:3.3.2）

Word Tips

pre-bid conference	投标前会议
huge and complex procurement	庞大而复杂的采购
highly itemized task description	高度逐项化的任务描述

Zhang Tian: The meal is good and I'm full now.

赵康:嗯,看来你的胃口很好。让我们继续,我知道你有很多问题。

Zhang Tian: You're right. I still have several questions for IFB.

赵康:别犹豫了,请问吧!

Zhang Tian: OK. Is the pre-bid conference a must in IFB?

赵康:通常情况下,对于一个庞大而复杂的采购,会召开投标前会议。

Zhang Tian: Is the bid deposit necessary?

赵康:不,不是。

Zhang Tian: As for pricing and pricing formats, is there anything particular?

赵康:可以这么说,投标人必须提交价格信息,例如单价或小时价以及总价。

Zhang Tian: I see. The unit or hourly rates and total prices are stipulated. I was told that some contracts require highly itemized task descriptions. Can you give me an example?

赵康:没问题。例如,用于地板翻新的 IFB 可能需要去除地板表面的胶质、焦油和类似物质。

Zhang Tian: Understood. So highly itemized task description can also go like this: electrical polishing machine with scrub brush should be used for a better mirror finish.

赵康:没错,你的学习能力很强。

Zhang Tian: Don't build me up too much.

第四节 篇章口译

Passage 1（音频:3.4.1）

【在 A 公司负责工程项目招标的经理向已通过初选的投标方介绍该公司的工程标书。】

First of all, I'd like to congratulate all of you that, after our examination of your prequalification documents, your companies have been accepted to enter the next stage for tendering of the power station project. This is the tender document for the civil works and you'll find there are many volumes.

Volume 1 is the Contract Conditions, which are worked out based on FIDIC Conditions. You know, FIDIC Conditions are fair to both contractors and clients, so we hope you will agree to these Conditions.

Volume 2 is the Bill of Quantities. There are 18 sections with 989 items of rates in the Volume. Please keep in mind that the contract price here is a rate basis contract, which means the total amount of the contract will depend on the actual quantities performed by the contractor. The structure of the power station is quite complicated and the quantities of the Works involve a great deal. Therefore, during the contract period, the rate for each item of the Works is fixed but the quantities of the Works will be modified accordingly.

Volume 3 is the Scope of the Works. You will find the description of each building and main structures of the power station. You should provide the detailed construction schedules for each building according to our general guiding schedule in this volume.

Volume 4 is Technical Specifications. The power station, which is quite different from ordinary buildings, requires many particular standards and methods. Since all the companies qualified to enter the next stage are experienced contractors for power stations, you should not feel strange with so many complicated specifications involved in this document.

Volume 5 is Quality Assurance. For this project, there are 3 grades of Quality

Assurance defined.

Volume 6 is Drawings, all of which are stamped by "For Tender Use Only".

Volume 7, the Bidding Documents, include Letter of Agreement, Construction Method Statement, Quality and Safety Programs and Bill of Quantities.

We'll make our decision according to our comprehensive analysis and judgment to your tender documents. We hope you will do your best to complete all the papers of the tender documents and hand it over to us within 30 calendar days. Good luck!

Passage 2 （音频：3.4.2）

【The head of the company's bidding department is explaining to the employees how to prepare for the submission of bids and the winning of bids.

公司招标部负责人正在向员工们介绍如何做好投标书提交和中标的准备。】

As described previously, Invitation for Bids (IFB) and Submission of Bids are two aspects of trade transaction. In fact, the roles of both purchaser and bidder are interchangeable. That is to say, a purchaser may become a bidder and a bidder may turn into a purchaser in various situations.

Since an IFB makes stringent requirements for the bidders, a winning bid proposal definitely takes time, knowledge, skill and effort. Any bid proposal with pitfalls can't be a competitive one and even the tiniest error may result in missing out on a profitable project. The following steps may be helpful for preparing a winning bid proposal and doing a good submission of bid.

Different projects have different requirements. From a practical point of view, you can't bid on every project you come across. Winning a contract that your company can't adequately perform can be just as costly as not winning it. Therefore, once you realize that your company won't make a reasonable profit if you are awarded the contract, you'd better dump it and move on to the next one.

Once you make your decision to submit a bid, you need to use your skills to find out everything you can about the company issuing the IFB, such as its market, organization, team and other affairs. After that, you may try to contact a prospective purchaser to gain some other relevant insights. Of course, you need to visit the site and attend the pre-bid meetings, if any, because these activities may help you do a better tender.

If you want to win a bid, you can't underestimate the importance of a tender. If you want to make your tender arresting, you need to group a professional team to prepare the tender elaborately on the basis of all stipulations and requirements of the IFB. You should follow strictly the tender's instructions, response forms which set strict word limits and ask questions geared towards seeking ideal responders.

A tender is so complicated that you need to re-check it throughout before the submission. Those factors, such as your quotation, references, signature, spelling, grammar, design and even aesthetic, are very important and need to be checked and confirmed. If it is possible, you may resort to some trustworthy friends to evaluate your tender. Once some errors are pointed out in your tender, you have your team members revise them in time.

Don't delay in tender submission. Any late tender is highly possible to be refused by the company or organization issuing the IFB. Take all factors into consideration, such as delivery, computer and Internet issues, and make sure that your tender can reach the right person or department on time. You can even consider handing in your tender in person if it is allowed.

To sum up, preparing submission of bids is not an easy job. If you want to win a contract, the five steps above may give you help.

第五节　口译技能——信息分层

一、训练目标

本单元在口译听辨的基础上对信息进行更深层次的分析,重点讨论一项信息处理技能——信息分层。

信息分层分为两大方面:第一,根据主要观点将发言内容划分为不同的部分;第二,区分各部分信息的价值,使结构层次分明。

通过英文语篇的概述练习,可以训练学生听辨时紧跟发言人的思路,梳理源语结构,捕捉语篇信息框架并区分信息价值的能力,为进一步练习打好基础。

二、练习技巧

(一)信息分层包括两个维度

1.根据发言的主要观点将信息划分为不同的部分。既然语篇内的信息是有机组合的,那么划分的依据可以是明显的语篇逻辑连接词,可以是过渡句或段,也可以是译员对语篇完整意义的整体理解与把握。然而,值得注意的是,并非所有发言人都拥有逻辑清晰的头脑和表达,有时语篇划分依据会不明显,这就需要译员对信息做重组。在中文发言中,这种特点会更加明显。

2.区分各部分信息的价值,使结构层次分明。当我们听到议论性发言时,中心观点就是主要意图。分论点就是主要信息,每一个分论点都有相应的论据,这些会被视为次要信息。如果发言者用了大量细节甚至故事去说明,这些信息的价值将进一步降低。例如,在议论性发言中,发言意图就是中心论点,主要信息就是分论点,每一个分论点都有论据,这些是次要信息。低级信息单位的重要性很大程度上取决于相应的高级信息单位。换言之,没有主要信息,次要信息就会支离破碎;没有次要信息,第三层级的信息也毫无意义。

3.结构清晰的发言有着明确的文本线索,不同观点之间也有显性的逻辑关系。而在模糊结构发言中,各类观点的边界没有用连接词标记出来,所以我们很难弄清不同观点之间该如何过渡。

比起英文发言,中文发言更容易出现模糊结构的问题,这可能有两个原因:

其一,中文是一种读者责任制的语言,理解信息的重担更多落在读者和听众身上。大家可以回忆一下熟悉的中国俗语,听者需要很费力才能弄清作者和发言者的意图。英语是作者责任制的语言,发言者有逻辑地表达观点,明示观点之间的联系。

其二,中文偏向归纳推理,英文偏向演绎推理。在演绎推理中,发言人通常从一个中心论点开始,并运用大量分论点和例证,翻译者便可以利用这种特性明确方向,确保思路正确。而在归纳推理中,我们首先会听到细节和例子,或是背景知识,最终才是发言人所总结的一般规律。

基于模糊结构的发言的特性,我们在具体翻译时可以使用以下两类小技巧:

第一,借助发言人的意图来预测结论。我们用粗略的预测来反驳或肯定获得的信息,经过预测,我们就可以更好地剖析发言中的每一个观点。

第二,在绘制思维结构图时,要有意识地明确观点之间的联系。即使发言人在发言中没有表达清楚,我们也要在大脑中"标出"逻辑关系。

三、示例精讲（音频:3.5）

以下材料选自英国萨塞克斯大学吉姆·沃森教授在 2008 年"首届中英低碳经济与政策国际论坛"上的发言,他的发言主题为" Policies to Support Low Carbon Innovation: National Strategies and International Cooperation"。请听材料并进行源语概述,听材料时不要参考发言原文。

So, what role for government? This is showing with you, really, a rather British-based perspective or a developed country perspective. But I see particularly from some of the things Professor Xia said, that actually some of the things are also discussed here. So the first two points, on this slide are the traditional ways of saying why government should intervene to support new technologies. One is to say that, if you leave firms to innovate on their own, they'll under-invest in research and development. That's a recognized so-called market failure in economics. And so government should come and compensate. And the reason is, the private firms cannot always capture the benefit from doing research themselves, so they will invest less than society needs them to do. The second reason is environmental externalities, such as air quality, earth deterioration, water pollution, or climate change. Again, that is a social problem, the private firms on their own will not tackle, so government needs to intervene to include this cost into economic decisions, through mechanism like carbon pricing, err, pollution fees and so on.

The second two points are additional points which come from study of successful environmental innovation. The first, you may seem obvious, and they just... not just apply to the environment, which is the need to build capacity and skills in new areas of technology, because people and firms don't suddenly appear fully formed, with these fully formed skills to implement these technologies. Sometimes government has a role to invest and help those skills to be developed. So, in the UK, you might see the need for, more firms that can install solar water heating for example, or another area is electricity networks. The companies that maintain our electricity networks in UK don't have very much capacity anymore in technology and innovation. So again, policy has to think about how it builds that technology capacity backup, if you want these networks to be innovative, and do different things to what they've done in the past.

The last point is lock-in, and that's often ex... ex... explored and, and discussed, which is that existing technologies and technology system that supplies us for example with energy, may have served us very well in the past in developed countries to give us reliable power for example, but may not be very good at, emm... limiting the amount of greenhouse gases we emit. And so, but, at the moment, firms and institutions and government regulations are locked into particular patterns of incentives and behavior. So government may need to intervene very strongly to change those, emm, so may, hopefully have to intervene more strongly than it would from conventional economic ideas.

第六节　课后练习

(音频:3.6)

I. Technical Terms Interpretation

1. 招标文件

2. 竞争性招标

3. 中标通知书

4. 货币要求

5. 现场考察

6. 信用记录

7. 建筑施工合同

8. 银行私人账户

9. 投标担保

10. 履约担保

11. 注册会计师

12. 公共集资建筑工程

13. award of the contract

14. bid invitation letter

15. original tender

16. period of validity of tenders

17. appendix to tender

18. tender submission date

19. specialized subcontractors

20. standards of workmanship

21. supplementary drawings

22. general provisions

23. legal litigation

24. arbitration

25. certified true copy

26. Assets & Debts Report

27. Profit & Loss Report

Ⅱ. Sentences Interpretation

1. 投标保证金是为了保护招标机构和买方不因投标人的行为而蒙受损失。招标机构和买方在因投标人的行为受到损害时可根据本须知第 15.7 条的规定没收投标人的投标保证金。十分感谢你方正式邀请我方参加上述工程项目的投标。

2. 他们承诺在投标人须知要求的日期之前提交投标书。

3. 我方已收到全部招标文件,正在准备投标书。

4. 投标人应提交金额为不少于投标报价总价2%的投标保证金,并作为其投标的一部分。

5. 很遗憾由于我方工程任务十分繁重,此次不能参加投标,敬请见谅。

Ⅲ. Dialogue Interpretation

A:打扰了,我可以在这领到招标文件吗?

B:Was your company qualified in the prequalification?

A:通过了。2 天前您的办公室已通知我们了。

B:Which company are you from?

A:中国建筑集团。

B:Please show me your company authorization letter for collecting the tender document and your name card.

A:给您。

B:Okay, that's fine. The tender document includes three booklets and thirty-six sheets of tender drawing. This is Booklet 1; it includes Instructions to the Tenderers, Form of Tender, Articles of Agreement, General Conditions of Contract and Special Conditions of Contract.

A:对履约保函有什么要求?

B:10% of the Contract Sum. The main points of concern in the Conditions of Contract, such as Construction Duration, Liquidated Damages, Percentage of Retention and Maintenance Period are abstracted in Appendix A to Form of Tender. Form of Performance Bond is shown in Appendix B to Form of Tender.

A:我明白了。那第二分册和第三分册呢?

B:Booklet 2 contains all the Specifications.

A:装修做法表也在第二分册中吗?

B:Yes, you can find it in the chapter for architectural works. Method of Measurement and Bill of Quantities are included in Booklet 3. This is a list of tender drawings. You can check the tender drawings against the list.

A:正好 36 张,没错。

B:Please sign here for my record.

Ⅳ. Passage Interpretation

【The manager in charge of the bidding of the project explains the precautions for the preparation of the tender documents to the bidders.

负责工程项目招标事宜的经理向各投标单位说明准备标书的注意事项。】

大家好,请注意! 请在准备标书时记得以下要求:

(1)投标者应按照规定准备一份标书原件和两个副本,并分别注明"原件"和"副本"。如果两者之间有不同,以原件为准。然后,投标者应把标书原件和两个副本分别装入一个内信封和一个外信封,且在信封上注明"原件""副本"。

(2)标书的原件和两个副本应打印或用不能抹掉的墨水书写,并由一名或多名有权使投标者遵守合同的人签字。此外,还应有一份用以证明授权的书面委托书与标书放一起。写有条目和有更正内容的每一页标书都要有标书签字人的签名。

(3)全套标书不应有改动、行间书写或涂抹的地方,按议程的指示或为改正投标者错误的情况除外,但这种情况下的更正应有标书签字人的签名。

(4)每位投标者只能提交一份标书,并且投标者针对一个合同只能投一次标。

第四章　工程合同谈判

第一节　背景知识

一、国际工程合同

(一)国际工程合同概念

合同是两个或两个以上当事人依法达成的明确双方权利和义务关系的具有法律约束力的协议。合同就是一种契约,在实践中,合同还会以不同的名称出现,如合约、合同书、协议、协议书、备忘录等。

国际工程合同是指不同国家的有关法人或个人之间,为了实现在某个工程项目中的特定目的,所签订的确定双方权利和义务关系的协议。由于国际工程是跨国的经济活动,因此国际工程合同远比一般国内合同复杂。

(二)国际工程合同特征

首先,具有国际性。国际工程合同的当事人是属于不同国家或地区的公司法人;承包人须在外国履行其全部或大部分合同义务,即合同的履行地是在承包人所属国以外的其他国家或地区,这会涉及复杂的法律适用问题。

其次,涉及巨额资金。国际工程合同标的通常很大,所以交易中通常需要各种各样的银行担保。

再次,需要长期履约。一般来说,国际工程合同期限长,需长期履约。

最后,承担较大风险。为减少风险,国际工程合同中除货币条款和不可抗力条款外,通常还含有价格调整条款、艰难情势条款等。

(三)国际工程合同分类

按工作范围分类,国际工程合同可分为全过程总包合同、设计—管理承包合同、设计—采购承包合同、管理承包合同、设计服务合同、咨询服务合同、施工或供货安装承包合同等。

按计价方式分类,国际工程合同可分为固定总价合同、固定单价合同、成本加酬金合同等。

按合同关系分类,国际工程合同一般分为总承包合同和分包合同,分包合同又包括施工分包、设计分包、劳务分包、设备材料供应分包、运输分包等。

(四)工程承包合同基本内容

招标成交的国际工程承包合同不是采取单一合同方式,而是由一系列的文件共同构成,统称为合同文件。合同文件包括招标通知书、投标须知、合同条件、投标书、中标通知书和协议书等。其中合同条件,按照国际惯例,一般包括监理工程师和监理工程师代表权责条款、工程承包的转让和分包条款、承包人一般义务条款、特殊自然条件和人为障碍条款、竣工和推迟竣工条款、专利权和专有技术条款、维修条款、工程变更条款、支付条款和违约惩罚条款等。

二、国际工程合同谈判

国际工程合同谈判,是弥补投标时由于时间有限、所掌握的资料有限而可能出现的差错的最佳时机,是承包商通过合同取得理想经济效益的关键一环。

合同谈判阶段实际上是在有关法律部门的监督下,在尽量遵守招标文件的前提下,业主和承包商为争取各自利益的讨价还价过程。从性质上讲,是指不同国家的有关法人为了实现在某个工程项目中的特定目的而签订的确定双方权利和义务的协议。合同一经签订,即具有法律约束力,任何一方不履行合同或履行不当,都要承担相应的法律责任。为了正确履行合同,双方必须在合同签订之前,熟悉合同条款,并把潜在的可能对一方不利的条款进行约定,分清不可预见事件的责任。

对于国际工程承包来说,业主通过初步评标并确认投标者的中标资格后,接下来便是艰苦的合同谈判阶段。每个国际工程承包项目的合同谈判从开始到结束的步骤各有不同,即使是相似的项目合同,完成谈判也不一定必须采取同样的步骤,谈判者必须保持最大的灵活性。

(一)谈判准备阶段

开始谈判前,一定要做好谈判的准备工作,只有这样才能在谈判中争取主动。谈判的准备工作包括如下几个方面的内容。

1.谈判的组织准备

包括组成谈判小组并选定谈判组长。一般来说,谈判组成员应包括有一定的法律知识、熟悉合同条款的商务人员,经验丰富的技术人员,熟悉当地情况的翻译人员和有着较为丰富的谈判经验、能驾驭整个谈判过程的谈判小组组长。

另外,在每一次合同谈判前,承包人的财务和法律方面的人员也应参与准备工作,但可根据具体情况决定是否实际参加谈判。谈判小组要充分发挥每一个成员的作用,组长能够协调组内工作。谈判组长应有较强的业务能力和应变能力,而且应该具备参加国际工程合同谈判的经验。谈判前须进行内部分工,确定谈判角色,以便在谈判桌上做到角色分明、相互配合、各有重点、进退自如。人员的大致分工为:

(1)商务人员:负责一般合同条款、特别行政条款、招标行政资料的研究。

(2)技术人员:阅读和熟悉当地技术规范和针对本工程的特殊技术规范。

(3)翻译人员:负责当地有关信息的收集,了解当地市场、项目特殊情况等。

(4)组长:负责对业主的要求给予回应,并担任谈判发言人。

2.谈判的方案准备

(1)开始谈判前,谈判小组需要认真研究所有的招标资料,列出需要解决的问题的清单,并根据具体问题拟定明确的解决方案以及回复对方方案的办法,写出谈判大纲,确定谈判的目标、任务和要求,将这一系列文件作为谈判工作的指导文件。

(2)了解对方的谈判人员以及他们的身份、地位、性格、爱好、办事作风,分析各自的优势和劣势。

(3)设计和确定最优方案、次优方案和备选方案,做到临场不乱。

(4)开始谈什么,接着谈什么,最后谈什么,事先都要有一个大致的安排。同时,要预计哪些环节可能出现分歧,出现了分歧应采取什么对策。

3.谈判的内容准备

国际工程承包合同的内容按优先顺序一般包括合同协议书、中标通知书、投标书和投标书附录、专用合同条件、通用合同条件、特殊技术规范、国家规范、图纸、标价的工程量清单、投标书附录中所列的其他文件等。谈判的内容准备主要应注意以下几个方面:

(1)招标文件中的投标人须知部分。在投标人须知中,业主会对合同范围、资金来源、对承包人的要求、标书文件的组成、评标办法等进行规定。很多人认为,这些内容只是对投标的指示,由于很少涉及合同具体内容而不被重视。但实际上,有时投标人须知中会隐藏有关合同实施的重要条款。

(2)合同条件部分。使用世界银行、非洲发展银行、亚洲开发银行等国际金融组织的资金的合同项目,通常使用 FIDIC 合同条件或其他国际通用合同条件,这些条款对所有投标人的要求相同,承包人不能改变。因此,此类合同谈判的重点是业主为此编制的合同专用条件,如付款方式、付款期间、质保金扣除比例与返还时间、业主风险等。对于业主自己编制的合同条件,则应仔细审查每一个条款。

(3)技术规范部分。对于合同实施地所在国的规范,我们不能改动,但可尽量争取在同等条件下使用我们熟悉的中国标准和规范;技术规范部分主要是看业主针对本项目编制的特殊规范,看其有无特殊要求,以及对我方不利或我方可以利用的规定。

(4)当地市场和项目的特殊性。不同的市场,或同一市场的不同项目,都有特殊性或特殊要求。合同谈判时,一定要指出这些特殊要求,减少合同执行过程中的麻烦。

(二)谈判进行阶段

1.谈判开始

此时主要是了解对方的基本情况,例如对方主谈人员有多大的决定权,是否还有幕后决策人,主谈人员的谈判风格和谈判策略等。除主谈人员外,其他谈判人员的情况也应注意,注意对方谈判人员的分工。

2.谈判中

谈判中应该向对方清楚表达自己的立场,不能因害怕谈判失败而回避自己的观点。谈判中双方都希望讨论自己关心的问题,在谈判中应注意引导对方转向自己关注的问题。谈判中出现僵局是常见的事,在这种情况下应该努力控制自己的情绪,并认真对待对方的观点,冷静分析其合理性,在对方要求合理的情况下,应该积极努力寻求共识。当然,不能轻易让步,即使对方要求合理。

3.表述我方意见

一旦提出不同的看法,就要论证自己立场的科学性和正确性,就要说明自己意见的事实依据或说明自己的意见符合国际惯例。要向对方讲明采纳意见后的利弊得失,谈判的目的无非是获利,如果对方感到有利可图,或者觉得不会失去利益,或者能更少地失去利益,就会十分重视这种意见,就更容易接受建议。在说服对方时,也应该坦率地说明自己的利益,使对方认为我方所提要求合情合理。谈判时要更多地强调双方利益的一致性。

4.反驳对方观点

一般说来,在准备阶段设计谈判程序时,就要努力避免辩论。但为了证明自己的立场,为了维护自身的合理要求,有时也不得不进行辩论。

(1)反驳时,可指出对方论点不正确,不符合国际惯例,或不符合原合同的规定;可指出对方的论据不可靠,或是不充分,或是根本就没有事实根据。

(2)采取原则问题不妥协、枝节问题不纠缠的方法。要抓住重点,切中要害。至于小问题,能妥协就妥协。

(3)措辞准确、犀利,但不要伤害对方,特别是不要刻薄讽刺,也不能断章取义,更不能歪曲对方的原意,特别是不要蛮不讲理。

(4)态度要客观、公正,要从容不迫、有条不紊、仪表庄重、举止自然。

(5)辩论的目的是较好地取得合同,因而应该有原则、有分寸、态度和善。

(三)合同签订阶段

谈判的结果由双方协商一致后形成的合同来体现。合同条款实质上反映了各方的权利和义务,合同条款的严密性与准确性是保障谈判获得各种利益的重要前提。因此,在达成一致意见后,拟定合同条款时,一定要注意合同条款是否完整、严密、准确、合理、合法,不要掉以轻心。防止被谈判对手在条款措辞或表述技巧上引入另一个陷阱,防止已经取得的结果、争取到的利益丧失殆尽。

第二节　相关词语与表达

序号	中文	英文
1	协议	agreement
2	一揽子项目	all-in-one program
3	一揽子合同	blanket contract
4	一揽子采购合同	blanket purchase contract
5	租船合同	charter contract
6	动产抵押	chattel mortgage
7	合同协议书	contract agreement
8	补偿贸易合同	contract for compensation trade

续表

序号	中文	英文
9	施工合同	contract for construction
10	咨询服务合同	contract for consulting services
11	合作开发自然资源合同	contract for cooperative development of natural resources
12	设计—施工合同	contract for design and build
13	设计—采购施工合同	contract for EPC
14	工程项目承包合同	contract for engineering project
15	合资经营企业合同	contract for joint venture
16	包工包料合同	contract for labor and materials
17	生产设备和设计—施工合同	contract for plant and design-build
18	项目合同	project contract
19	货物销售合同	contract for sale of goods
20	技术转让合同	contract for technology transfer
21	交钥匙工程合同	contract for turnkey project
22	包工合同	contract for work
23	协作合同,联合合同	contract of association
24	运输合同	contract of carriage
25	海运合同	contract of carriage by sea
26	雇佣合同	contract of employment
27	雇工合同	contract of hire of labor
28	保险合同	contract of insurance
29	租赁合同	contract of lease
30	管理合同	management contract
31	指定的分包合同	nominated sub-contract
32	竞业禁止协议	non-compete agreement
33	一揽子交易	package deal
34	勘察合同	survey contract
35	房地产契约	title deed

续表

序号	中文	英文
36	裁决员	adjudicator
37	借款人,借用人	borrower
38	买方,业主,受托人,顾客	buyer/client
39	承包商,承包人	contractor
40	承包商(人)代理人	contractor's agent
41	承包商人员	contractor's personnel
42	承包商(人)代表	contractor's representative
43	协调员	coordinator
44	争端裁决委员会	dispute adjudication board(DAB)
45	雇主(业主)代表	employer's representative
46	工程师	engineer
47	工程师代表	engineer's representative
48	总承包商	general contractor
49	联营体,合资企业	joint venture（JV）
50	地主,房主	landlord
51	贷款方,出借人	lender
52	承租人,租地人	leaseholder
53	领到执照的人,获许可的人	licensee
54	许可证颁发人,出证人,出让方	licensor
55	抵押贷款人	mortgagee
56	甲方	party A
57	乙方	party B
58	签订协议(合同)的各方	parties to the agreement（contract）
59	买方	purchaser
60	卖方	seller
61	分包商	sub-contractor
62	供货商	supplier
63	承租人	tenant
64	(甲、乙)双方	the two parties
65	买方	vendee

续表

序号	中文	英文
66	合同条件	conditions of contract
67	定义条款	definition provision
68	陈述与保证条款	representation and warranties provision
69	补偿条款	indemnification provision
70	保密条款	confidentiality provision
71	终止条款	termination provision
72	合同的转让与变更条款	assignment and modification provision
73	不放弃条款	non-waiver provision
74	不可抗力条款	force majeure provision
75	竞业禁止条款	non-competition provision
76	知识产权条款	intellectual property right provision
77	违约条款	default provision
78	纠纷解决条款	disputes resolution provision
79	先决条件	condition precedent
80	保险条款	insurance provision
81	税收条款	taxation provision
82	独立合同关系条款	independent contract relationship provision
83	完整性条款	whole agreement merger provision
84	日历日(月,年)	calendar day (month, year)
85	条款	clause
86	特许协议	concession agreement
87	土木工程建筑合同条件	conditions of contract for works of civil engineering construction
88	特殊应用条件	conditions of particular application
89	转包,分包	to sub-let
90	合同安排	contractual arrangement
91	合同号	contract number
92	合同期	contract period
93	合同价格	contract price
94	合同项目	contract project

续表

序号	中文	英文
95	合同风险	contract risks
96	合同序号	contract serial number
97	合同签署日期	contract signature date
98	合同条款	contract terms
99	承包商的责任	contractor's liability
100	定义	definition
101	菲迪克(国际咨询工程师联合会)	FIDIC
102	不可抗力	force majeure
103	通用条件,一般条件	general conditions
104	公历,阳历	Gregorian calendar
105	违约责任	liability of breach of contracts
106	其他条款	miscellaneous clauses
107	专用条款	particular conditions
108	支付条款	payment provisions
109	保留条件	proviso
110	保留条款	proviso (reservation, saving) clause
111	特别规定,特殊条款	special provisions
112	子条款	sub-clause
113	违反合同	to violate the contract
114	付款条件	terms of payment
115	有效合同	valid contract
116	无效合同	void contract
117	签订和履行合同	signing and performing contracts
118	违约,违反合同	breach of contract
119	承包商违约	breach of contractor
120	合同的执行	contract performance
121	合同记录	contract records
122	合同续订	contract renewal
123	合同范围变更	contract scope changes
124	合同暂停	contract suspension

续表

序号	中文	英文
125	合同终止	contract termination
126	合同变更	contract variation
127	承包商的许诺	contractor's undertaking
128	对业主的合同承诺	contractual commitments with the client
129	与……签订合同	(the) entry of a contract with
130	定期会议	periodic meeting
131	执行范围	scope of execution
132	中止合同	suspension of contract
133	终止合同	termination of the contract
134	遵守合同	to abide by the contract
135	违反合同	to break the contract
136	撤销合同	to cancel the contract
137	执行合同,履行合同	to carry out the contract
138	改变合同	to change the contract
139	与某人达成框架协议	to come to a framework agreement with sb.
140	遵守合同	to comply with the contract
141	签订临时协议	to conclude a provisional/temporary agreement
142	起草合同	to draft a contract
143	拟定合同	to draw up a contract
144	与……订立(或签订)合同	to enter into a contract with
145	签署合同	to execute a contract
146	未能与某人就某事达成协议	to fail to reach an agreement with sb.
147	履行义务	to fulfill an obligation
148	履行合同	to perform the contract
149	签订合同	to sign a contract
150	推卸责任	to skirt one's responsibility

第三节　对话口译

Dialogue:采购物资（音频:4.3.1）

Word Tips

consume	消耗,消费
construction period	施工期
draft	草案,起草
clause	条款
amend	修正,改正
finalize	最终确定
diesel	柴油
engine oil	机油
hydraulic oil	液压油
lubricating oil	润滑油

【场景:To keep all the vehicles and equipment running on the job site, a large quantity of fuels and oils are consumed every day. Mr. Niko, company represent of CCCC (China Communications Construction Company) is trying to negotiate a supply contract with Ms. Nancy, sales manager of Esso Standard Oil Ltd., for the whole construction period.

要保持现场所有车辆与设备的运行,每天都需要消耗大量燃料和其他非燃料用油。中国交建公司代表 Niko 先生正在努力与埃索标准石油有限公司的销售经理 Nancy 女士谈判一个贯穿整个施工期的供油合同。】

Niko: Ms. Nancy, we have examined the draft of the supply contract prepared by your company. There are some clauses that I think need to be clarified or amended.

Nancy: 那么,我们一条一条地讨论好吗? 看看今天能不能定下来。

Niko: OK. Let's start with Clause 2 "Object of Contract". In addition to premium gasoline and diesel, would you be able to supply oils as well? You know, we also need a lot of them for the operation of our equipment.

Nancy: 埃索公司很乐意供应各类非燃料用油,如机油、液压油和润滑油等。

Niko: That's good. Now, let's move on to Clause 4. It's about the term of the supply contract. In this clause, it says, "The term of the Supply Contract is three years from this date and the Contract will expire automatically on the first day of January 2026." It's true that the completion time of the Highway Project will be three years according to our present construction schedule. However, the Project could be delayed for a period of time or completed ahead of schedule. You never can tell, even an experienced contractor like us. Can we change the wording to this: "The term of the Supply Contract starts on the first day of January of 2023 and will expire on the date of completion of the Highway Project."

Nancy: 我没有意见。

Niko: In Paragraph 3 of Clause 6, which states the manner of delivery, does the term "the Buyer's order" include both verbal and written orders?

Nancy: 我们的意图是"买方的订单"指的是书面的,不包括口头订单。

Niko: I hope you will accept our verbal orders, too. Sometimes we are in urgent need of some items, which allows us no time to send you a written order.

Nancy: Niko 先生,我们愿意尽力为你方提供各种便利,但口头订单很容易出问题。

Niko: Why not put the wording this way: "…after having received from the Buyer a written order, or a verbal order in which case a letter of confirmation will follow accordingly."

Nancy: 这我可以接受。

第四节　篇章口译

Passage 1:业主代表就合同细节发言（音频:4.4.1）
Word Tips

large and durable equipment	大型耐用设备
easily consumable machine	易损机械
concrete pump	混凝土泵

concrete batching plant	混凝土搅拌站
concrete mixing truck	混凝土搅拌车
crane	吊车
construction equipment list	施工设备清单
mobilize	调动
schedule of equipment mobilization	机械设备进场计划表
construction schedule	施工计划表
idling of equipment	机械设备闲置
cost breakdown	成本细分清单
sub-contractor	分包商
prestressing works	预应力工程
cladding and roofing works	墙围护和屋面工程
HVAC(Heating, ventilation and Air Condition)works	暖通空调工程
lump sum price	包干价
rebar	钢筋
formworks	模板
Bill of Quantities	工程量表

【场景：承包人与业主就合同细节展开谈判，以下是业主代表的发言。】

We've noticed that the total cost of construction equipment is USD 8.01 million, which is really too high. We know something about the depreciation periods for common equipment. Generally speaking, for large and durable equipment, it is 15 years. The depreciation period is 10 years for some ordinary equipment and 3 years for some easily consumable machines. As you have mentioned that this is a large project with a tight program, much high efficient equipment such as concrete pumps, high capacity concrete batching plants, concrete mixing trucks and many kinds of cranes would be employed which are shown in our construction equipment list.

We totally understand that you should mobilize enough equipment for the construction, but in order to control the machinery cost, you'd better prepare a schedule of equipment mobilization, which should closely match the construction schedule to avoid any idling of equipment on site. You also need to check the equipment cost and see if there are any costs that can be deducted.

What's more, the cost for labor and staff camp is too high. Your budget for the temporary house is USD 60/m². It is unnecessary because the local workers can use the house made by a kind of local plant which costs USD 35/m² only.

So, we need a detailed cost breakdown for the camp which is divided into two parts: for staff and for labor. I do not think that the cost for staff living area will exceed USD 60/m² and the cost for labor USD 40/m². In this case, there is a big amount of price which can be deducted.

Item 6 is the cost for major sub-contractors. You mentioned sub-contractors for the prestressing works, cladding and roofing, and HVAC (Heating, ventilation and Air Condition) works. However, we can't accept the cost for these special works submitted to us by just several lump sum prices. You are requested to provide us with the same detailed analysis as you did on concrete, rebar, formworks and others in the Bill of Quantities. Otherwise the cost breakdown is not a complete one.

Passage 2:承包人代表发言（音频:4.4.2）

Word Tips

commence the works	开工
extension of time to complete the works	延期完工
key date	关键日期
delay of completion	完工延误
retention period	保修期
retention money	保修金
completion certificate	完工证书

【场景:承包人与业主就合同细节展开谈判,以下是承包人代表的发言。】

对于本合同的条件,我们有如下意见,希望你们予以考虑:

第三十条规定,承包人必须在签订合同后的 10 天内开工。我们认为这有点紧张。这是一个大工程,我们需要更长的时间来为开工做准备。因此,我们建议将"10 天"修改为"15 天"。

第三十六条只提及承包人必须做的事,而没提及现场未能在双方同意的时间移交给承包人的情况下,承包人的权利。因此,我们认为有必要加上这样一条:如果承包人不能按时得到场地或因占有场地或部分场地而增加了费用,那么承包人有获得工程完工日期延期和合同价格加价的权利。

我们不同意第四十三条中关于误期违约金的内容。首先,关键日期每级的违约金总额太高。例如,你们规定,四星级关键日期的违约金为每天20000美元,这远远高于通常的金额。去年我们完成了一个类似的工程,合同条件中支付的违约金只有这个数目的一半。另外,既然针对完工延误有惩罚,那么我们在关键日期之前提前完工应当有奖金。

我们对第五十条中提及的保修期和保修金有不同的看法。本条款提及保修期为整个工程完工证书发布后的一年。我们认为计算保修期应从分项工程的完工证明签发的那天开始计算才合理,因为各分项工程完工之间的间隔时间很长。

第五节　口译技能——合同谈判中的口译技巧

一、训练目标

国际工程谈判的基本目标是进行交流,达成某种结果——追求利益最大化。口译者应尽力消除谈判者间的理解障碍,促成工程项目的成功。

国际工程项目谈判通常又称大型项目谈判,例如利用外国政府或国际金融组织的贷款,对一些大型市政建设和环保项目以及重要的技术改造项目进行的谈判。这种谈判主要围绕项目的目的、内容、发展前景、融资条件、招标与发包等一系列经济与技术上的问题进行谈判。

建设项目的谈判通常分两部分进行,第一部分是由双方政府主管该项目的部门会同有关经济部门就双方合作的总体设想和商务关系进行的原则谈判,谈判涉及面较广,包括建设项目的性质、作用,建设项目的投资、贷款总额及支付方式,建设项目建设过程中双方的权利、责任等。第二部分是具体的技术和商务谈判,由双方的具体实施建设工程的部门或企业直接谈判,谈判涉及的内容较专业,往往就其中一些技术细节和工程所用材料和设备、工程的技术标准、验收方式等进行谈判。前后两部分谈判是相辅相成的,第一部分的谈判决定了第二部分谈判的范围和要求,而第二部分谈判也对第一部分的谈判做了必要的补充。两部分的有机结合和互相补充决定了整个建设项目的成败,所以这种谈判要比其他谈判更加复杂,要求也更高。

二、练习技巧

1. 合理使用模糊语言

(1) I think it's hard for our clients to accept the quality of your construction work. At this price and quality, we can't play in the same ball park.

我认为我方客户很难接受贵方的建筑工程的质量。以这个价格和质量,我们是谈不到一块了。

(2) We have always insisted on the principle of "equality and mutual benefit and exchange of needed goods", and we have adopted much more flexible methods in our dealings nowadays.

我们一直坚持"平等互利,互通有无"的原则,而且我们现在的做法比以前灵活多了。

在上述句子中,诸如"much more flexible methods"(更加灵活的方法),都是笼统的表达,向别人传递模糊的信息,间接地表达自己的想法。这些模糊表达法适用于直接语言不宜使用的场合,能收到更好的谈判效果。

2. 正确使用逻辑语言

谈判的目的是磋商工程质量以及工程实施细节,为签署工程合同做准备。谈判涉及工程项目甲乙双方的权利和义务,所以语言必须准确、规范,具有很强的逻辑性,因为它实际上是一种法律语言。

We believe that you should issue us a certificate of excellent performance.

我方认为贵方应当给我方颁发优秀履职证书。

上句用词不当,略显生硬,应改译为:We deserve a certificate of excellent performance.

This contract will come into effect from Oct. 1.　　本合同从 10 月 1 日起生效。

上句没有说明 10 月 1 日是否包括在内,应改译为:

This contract will come into effect from and including Oct. 1, 2004.

本合同从 2004 年 10 月 1 日(包括 10 月 1 日在内)起生效。

3. 恰当使用礼貌语言

在谈判中发言时,很重要的一点就是要掌握好交际原则。避免诸如"our

esteemed clients"" your favor"" your valued order"之类的陈腐词语和商业行语,以及"permit""allow""extremely and sincerely sorry"等通常用以表示卑恭之意的词语来向别人表达你的要求。

三、实例精讲

在国际工程项目谈判的不同阶段,谈判者有不同的交际主题,想达到的目的也不同,因此每个阶段的口译策略也可能不同。

1. 初始阶段

在这个阶段,双方想去了解对方,从而营造一种和谐的谈判气氛。口译策略必须为这个总目的服务。例如:双方在谈判开始时互赠礼物以示友好,中方代表给客人送上礼物时一般会说:"这是我们的小礼物,不成敬意,请笑纳。"在分析了讲话者的意图和接受者的期待功能后,口译者对源语做出修改以表达出说话人的交际意图和真正目的。因此,中方代表的话被译成了"Now, we have some presents for you and hope you would like them. "。这个翻译与说话人的字面意义是不对等的,但这个翻译符合了双方在这个阶段的目的——互相了解,营造和谐的谈判气氛。这个翻译能够让对方接受,口译者有意的不忠实翻译比直译的效果好。

2. 谈判关键阶段

这个阶段是整个谈判过程中最重要的阶段,交际目的是促使谈判和交易成功。此阶段的口译必须达到这个总目的。例如:听完中方总经理最后的报价后,外方代表说"Unless you can reduce the price by 3%, there is no chance for the transaction. "。这句话的直译是:"除非贵方能降3%,否则这门生意没有成交机会。"在此情况下,口译者不能使用其他词语来改变源语的语气,否则就会与双方此时表明态度的目的相抵触。

另一种译法——"除非你们能降价3%,否则这个生意成交的机会就很小了"听起来更容易让人接受。但这种译法会给接收方不同的信息,引起误会。这时,口译者不需要考虑双方言语的礼貌问题或双方言语的后果,因为所用语言和语气都是谈判者有目的的行为,也是他们的谈判策略。所以在此种情况下,直译要比意译好。

3. 方案确定阶段

当进入了这个阶段,讨论了所有必要的工程条款后,谈判者会相互表示感谢或表达进一步合作的愿望。口译者的策略,不管是直译还是意译,都要为此目的服务。例如:1993 年美国某输电公司与中方某电厂谈判融资项目,美方想尽快派两名项目管理人员到中方监督电厂二期工程的建设,美方主谈讲道:"...(lay stress on the utterance) Is that a big deal? We know, your company can sure stretch a dollar. We have proven track record in joint ventures with other power plants in China. (in a sense of humor) We won't ship tons of people here!"美方聘请的译员用一种生硬、缓慢、正式的口吻译为:"我们觉得此事非同小可!我们知道贵公司会精打细算,但我们和中国其他电厂合作时也采取了这种方法,有证可查。我们又不是把好几吨人运过来!"中方主谈听完译文后神情严肃,认为美方太霸道,说话太冲,现场气氛顿时凝固了。虽然会后美方一再向中方解释,但最终双方在外派项目管理人员上没有达成任何协议。其实译文与美方主谈的真实意思有出入。如果译员把主谈这番话译为:"(加重语气)我方派项目管理人员来贵公司,对贵公司来说不算什么吧!我方知晓贵公司考虑到费用问题。我公司与中方电厂成立的合资公司都有很好的业绩,(诙谐的口吻)关于外派人员方面,我方又不是把成千上万的人运来!"效果会更好。

第六节　课后练习

(音频:4.6)

Ⅰ. Technical Terms Interpretation

1.合同条件

2.符合

3.预付款

4.合同价款

5.工程进度款

6.进场费用

7.履约担保

8.支付条款

9.付款条件

10. 有效合同

11. void contract

12. contract suspension

13. contract termination

14. to fulfill an obligation

15. to perform the contract

16. to violate the contract

17. to sub-let

18. contractor

19. contractor's agent

20. contractor's personnel

Ⅱ. Sentences Interpretation

1. 我们建议在 6.6 款的第一句话后加上一句话。

2. 合同将我们达成协议的内容基本都写上了。

3. 那么建设工程合同条件涉及哪些主要内容?

4. 一旦你们中标了这个项目,你们将会收到盖有"施工专用"标志的图纸。

5. 这是一种严重的违约行为,为此我方要求得到适当的补偿,我方保留为此要求赔偿的一切权利。

6. 如果一方不执行合同,另一方有权撤销该合同。

7. 你是否担心我们不履行合同或者拒不付款?

8. 请会签第 156 号销售合同一式两份中的一份,并将它寄回我方存档。

Ⅲ. Dialogue Interpretation

【场景:承包人与业主就合同细节展开谈判,以下是承包人代表张兴(简称Z)和业主代表 David(简称 D)之间的对话。】

D:Today we will enter the stage of negotiating the Contract. In this stage, we'll discuss both Contract Conditions and prices. According to the schedule, we'll spend a couple of days on each topic.

Z:很高兴能有机会和你们讨论一下合同条件。

D: Now let's come to the point. First, I'd like to make it clear that the Contract Conditions are in line with the general principles of FIDIC.

Z: 我们非常赞成采用 FIDIC, 因为这对业主和承包人都比较公平。

D: I'm sure you must have read these conditions carefully and now please tell us your comments.

Z: 坦率地讲, 如果你们不介意的话, 我们对这个合同条款确实有不同的看法。合同中没有关于支付给承包人的预付款的条款。通常情况下, 在签订合同后, 承包人应当得到大约 10% 的合同价款来进行工程准备。

D: I'm afraid that there is no room for discussion in this point. According to the Contract, the payment can be started only from the progress payment and you must prepare the fund for your mobilization cost by yourselves.

Z: 既然这样, 我们撤回这一意见。第八条是关于履约担保的, 我们认为 15% 的保证金比通常情况高。我们希望将其调整为 10%。

D: We'll consider what you proposed for this Clause.

Ⅳ. Passage Interpretation

Honorable Vice Minister, it is my privilege to meet with you today. We have five suggestions as follows. Firstly, we think some important information should be compiled into Concession Agreement as attachments. This information includes project site locations, site delivery schedule, technical requirements, operation and maintenance requirements, takeover procedure, financing structure, cash flow projection, government support agreement, etc. Secondly, we suggest simply merging 5 years' construction period and 25 years' O&M period into 30 years' concession period. The third suggestion is that the Concession Agreement prohibits competing road construction, including highway, railway or upgrading the parallel existing road which can divert revenue streams away from this project. The fourth suggestion is that the Concession Agreement should set out the compensation mechanism for low traffic flow rate. The last one we suggest is to have an indexation of the toll rate linked to CPI, and accordingly, the toll can be adjusted on an annual basis according to your country's inflation.

第五章　工程施工准备

　　施工准备工作的基本任务是为拟建工程的施工创造必要的技术和物质条件,统筹安排施工力量和施工现场。施工准备工作也是施工企业搞好目标管理、推行技术经济承包的重要依据。同时施工准备工作还是土建施工和设备安装顺利进行的根本保证。因此,认真地做好施工准备工作,对于发挥企业优势、合理供应资源、加快施工速度、提高工程质量、降低工程成本、增加企业经济效益、赢得企业社会信誉、实现企业管理现代化等具有重要的意义。

　　实践证明,凡是重视施工准备工作,积极为拟建工程创造一切施工条件的,其工程的施工就会顺利地进行;凡是不重视施工准备工作的,其工程的施工就会遇到麻烦或面临损失,甚至给工程施工带来灾难,其后果不堪设想。

第一节　背景知识

　　施工准备工作通常包括技术资料准备、施工物资准备、劳动组织准备、施工现场准备和施工现场外准备五个方面。

　　一、技术资料准备

　　技术资料准备是施工准备工作的核心,是确保工程质量、工期、施工安全和降低工程成本、增加企业经济效益的关键,因此必须认真地做好技术资料准备工作。其主要内容有熟悉与会审施工图纸、调查研究与收集资料、编制施工组织设计、编制施工预算文件。

　　(一)熟悉与会审施工图纸

　　1.熟悉与会审施工图纸的目的

　　(1)充分了解设计意图、结构构造特点、技术要求和质量标准,以免施工中发生指导性错误。

　　(2)通过审查发现设计图纸中存在的问题和错误,使其在施工作业之前改正，为施工项目的实施提供一份准确、齐全的设计图纸。

（3）提出合理的建议和协商有关配合施工等事宜，确保工程质量和安全，降低工程成本和缩短工期。

2.熟悉施工图纸需把握的内容和要求

（1）基础部分。核对建筑、结构、设备施工图纸中有关基础留洞的位置、尺寸、标高，地下室的排水方向，变形缝及人防进出口的做法，防水体系的做法和要求等。

（2）主体结构部分。掌握各层所用砂浆、混凝土的强度等级，墙、柱与轴线的关系，梁、柱配筋及节点的做法，悬挑结构的锚固要求，楼梯间的构造做法等，核对设备施工图和土建施工图上洞口的尺寸与位置关系是否准确、一致。

（3）屋面及装修部分。掌握屋面防水节点的做法，内外墙和地面等所用材料及做法，核对结构施工时为装修施工设置的预埋件、预留洞的位置、尺寸和数量是否正确。

在熟悉图纸时若发现问题，应在图纸的相应位置做出标记，并做好记录，以便在图纸会审时提出意见及解决问题的建议，协商解决。

3.施工图纸会审的重点内容

（1）审查拟建工程的地点、建筑总平面图同国家、城市或地区规划是否一致，与规划部门批准的工程项目规模、形式、平面立面图是否一致，在设计功能和使用要求上是否符合节能、环保、卫生、防火及美化城市等方面的要求。

（2）审查施工图纸与说明书在内容上是否一致，施工图纸是否完整、齐全，各种施工图纸之间或各组成部分之间是否有矛盾和误差，图纸上的尺寸、标高、坐标是否准确、一致。

（3）审查地上工程与地下工程、土建工程与设备安装工程、结构工程与装修工程等施工图纸之间是否有矛盾或施工中是否会互相干扰，地基处理、基础设计是否与拟建工程所在地点的水文、地质条件等相符合。

（4）当拟建工程采用特殊的施工方法和特定的技术措施，或工程复杂、施工难度大时，应审查本企业在技术上、装备条件上或特殊材料、构配件的加工订货上的力量或能力，能否满足工程质量、施工安全和工期的要求，采取某些方法和措施后，是否能达到设计要求。

（5）明确施工项目的结构形式和特点，复核主要承重结构的强度、刚度和稳定性是否满足要求,审查施工图纸中复杂、施工难度大和技术要求高的分部、分项工程或新结构、新材料、新工艺,检查现有施工技术水平和管理水平能否满足工期和质量要求,并采取可行的技术措施加以保证。

(6)明确建设期限、分期分批投产或交付使用的顺序、时间,以及建设单位提供的材料和设备的种类、规格、数量及供货日期等。

(7)明确建设、设计、监理和施工单位之间的协作、配合关系,以及建设单位可以提供的施工条件。

(二)调查研究与收集资料

通过实地调查研究,收集与施工项目相关的各类资料,以备使用。

(三)编制施工组织设计

施工组织设计是全面安排施工生产的技术经济文件,是指导施工的主要依据。施工总承包单位经过投标、中标承接施工任务后,即开始编制施工组织设计,这是拟建工程开工前最重要的施工准备工作之一。

(四)编制施工预算文件

施工组织设计经批准后,即可着手编制单位工程施工图预算和施工预算,以确定人工、材料和机械费用的支出,并确定人工数量、材料消耗数量及机械台班使用量。

二、施工物资准备

主要材料、地方材料、构配件、脚手架、施工机具等施工用的物资是确保拟建工程顺利施工的物质基础,这些物资的准备工作必须在工程开工前完成。根据各种物资的需要量计划,分别落实货源,安排运输和储备,使其满足连续施工的要求。

三、劳动组织准备

劳动组织准备是确保拟建工程能够优质、安全、低成本、高效地按期建成的必要条件。其主要内容包括:研究施工项目组织管理模式,组建施工项目经理部;建立精干的施工队伍;建立健全质量管理体系和各项管理制度;完善技术检测措施;落实分包单位,审查分包单位的资质,签订分包合同;加强职业培训,做好技术交底工作。

四、施工现场准备

施工现场是施工的全体参与者为优质、高效、低耗地完成施工目标,而有节奏、均衡、连续地施工的活动空间。

五、施工现场外的准备

施工准备除了施工现场内部的准备工作,还有施工现场外的准备工作。其具体内容如下:

1. 材料的加工和订货

建筑材料、构配件和建筑制品大部分必须外购,工艺设备更是如此。如何与加工部、生产单位联系,签订供货合同,搞好及时供应,对于施工企业的正常生产是非常重要的;对于协作项目也是这样,除了要签订议定书,还必须做大量的有关方面的工作。

2. 做好分包工作和签订分包合同

由于施工单位本身的力量所限,有些专业工程的施工、安装和运输等均需要委托外单位,根据工程量、完成日期、工程质量和工程造价等内容,与其他单位签订分包合同,保证按时施工。

3. 向上级提交开工申请报告

当材料的加工和订货已完成及做好分包工作和签订分包合同等施工现场外的准备工作后,应该及时地填写开工申请报告,并上报上级批准。

第二节　相关词语与表达

序号	中文	英文
1	施工	construction
2	施工安装图	construction and erection drawing
3	施工安装用的机械及工具	machines and tools for construction and erection
4	施工标桩	construction stake
5	施工布置图	construction plan
6	施工步骤	construction procedure
7	施工测量	construction survey
8	施工程序	construction program
9	施工程序网络图	project network

续表

序号	中文	英文
10	施工贷款	construction loan
11	施工单位	construction organization
12	施工地点	job location
13	施工吊车	construction hoist
14	施工定额	construction norm
15	施工队	construction team
16	施工方法	job practice
17	施工方式	form of construction work
18	施工费用	construction cost
19	施工缝	construction joint
20	施工工程	construction work
21	施工工程师	construction engineer
22	施工工程学	construction engineering
23	施工工期	construction period
24	施工工种	construction trade
25	施工规模	scope of construction item
26	施工荷载	construction loads；working load
27	施工合同	construction contract
28	施工机械	construction machinery
29	施工机械费	cost of constructor's mechanical plant
30	施工机械化	mechanization of building operation
31	施工机械化系数	coefficient of construction mechanization
32	施工技术财务计划	financial plan for construction technology
33	施工计划	construction plan；construction program
34	施工监督	supervision of construction
35	施工检查	inspection of construction
36	施工检查员	construction inspector
37	施工阶段	construction stage
38	施工进度	construction progress
39	施工进度表	schedule of construction

续表

序号	中文	英文
40	施工进度计划	construction schedule
41	施工经理	construction manager
42	施工卷扬机	builder's hoist; builder's winch
43	施工临时螺栓	construction bolt
44	施工流水作业法	construction streamline method
45	施工面积	floor space under construction
46	施工平面图	construction plan
47	施工期	construction period
48	施工企业	construction enterprises
49	施工前阶段	preconstruction stage
50	施工区	construction area
51	施工缺陷	constructional deficiency
52	施工人员	constructor
53	施工设备	construction equipment
54	施工水平仪	builder's level
55	施工说明	general description of construction
56	施工说明书	construction specifications
57	施工图设计	construction documents design
58	施工图设计阶段	construction documents design phase
59	施工图预算	working drawing estimate
60	施工文件	construction documents
61	施工误差	construction error
62	施工现场	fabricating yard
63	施工详图	detail of construction
64	施工项目	project under construction; construction item
65	施工项目编号	construction item reference number
66	施工效率	efficiency of construction
67	施工性能	workability
68	施工许可证	builder's license
69	施工验收技术规范	technical code for work and acceptance

续表

序号	中文	英文
70	施工依据	manufacture bases
71	施工预算	construction estimate
72	施工准备	preliminary work for construction
73	施工准备计划	preparatory plan
74	施工总平面图	overall construction site plan
75	施工总则	general conditions of construction
76	施工组织设计	construction management plan
77	施工作业计划	work element construction program
78	项目	item
79	项目报告	project paper
80	项目编制阶段	project preparation phase
81	项目标记	project mark
82	项目财务估价	project financial evaluation
83	项目采购	project purchasing
84	项目成本	project cost
85	项目筹备融通资金	project preparation facility（PPF）
86	项目贷款	project loan
87	项目单	menu
88	项目的拟定	project formulation
89	项目的所有者	owner of the project
90	项目的总投标价值	total tender value of project
91	项目发展周期	project development cycle
92	项目范围	scope of project
93	项目方案	project alternatives
94	项目分隔符	item separation symbol
95	项目分类	classification of items
96	项目分配	allocation of items
97	项目分析	item analysis
98	项目工程师	project engineer
99	项目工程师负责制	project engineer system

续表

序号	中文	英文
100	项目号	item number
101	项目核对法	check list
102	项目核准权	projects approval authority
103	项目后评价	post project evaluation
104	项目后评价和后继行动	post project evaluation and followup
105	项目划分	segregation of items
106	项目化组织	projectized organization
107	项目环境生态评价	environmental appraisal of a project
108	项目技术评价	project technology evaluation
109	项目计划	project plan
110	项目计划设计图	project planning chart
111	项目记录	item record
112	项目加权	item weighting
113	项目检验回归曲线	item-test regression curve
114	项目建设费用	cost of project implementation
115	项目建设进度表	project implementation scheduling
116	项目建议书	proposals for the projects
117	项目经济估价	project economic evaluation
118	项目经理	project manager
119	项目矩阵组织	project matrix organization
120	项目控制	project control
121	项目块	entry block
122	项目历史	history of project
123	项目目标管理	project management by objectives
124	项目资金筹措	project financing
125	项目匹配法	matching item
126	项目评估法	project evaluation and review technique
127	项目企业经济评价	enterprise's economic appraisal of a project
128	项目删除	deletion of items
129	项目设备	project equipment

续表

序号	中文	英文
130	项目设计	item design
131	项目设计工程师负责制	project engineer responsibility system
132	项目生产	project production
133	项目说明	item description
134	项目说明书	specification of an item
135	项目投资时期	investment phase of a project
136	项目图与布置图	project charts and layouts
137	项目网络法	project network technique
138	项目网络分析	project network analysis
139	项目文件	item file
140	项目修理通知单	item repairing order
141	项目一次采购法	project purchasing
142	项目预算编制	program budgeting
143	项目支助业务	project supporting services
144	项目执行	project implementation
145	项目执行情况审计报告	project performance audit report（PPAR）
146	项目指标	project indicators
147	项目转移	item advance

第三节　对话口译

Dialogue 1:讨论施工现场布置图（音频:5.3.1）

Word Tips

precast yard   预制构件场

gantry crane   龙门吊

rebar yard   钢筋场

rebar (bending) shop   钢筋车间

bending machine   (钢筋)折弯机

cutting machine	（钢筋）切断机
iron shop	铁件车间
batching plant	（混凝土）搅拌站
catch basin	集泥井

Mr. Armstrong: Hi, Mr. Bai!

白先生：您好，阿姆斯特朗先生！

Mr. Armstrong: Did you bring your proposal of the site layout with you?

白先生：带来了。我昨天晚上为准备它一直干到午夜。我知道你们很着急。

Mr. Armstrong: You know there are many parties working on the same site while the total site area is limited. You should arrange your site in a compact manner.

白先生：我明白。给您！预制构件场、钢筋场、混凝土搅拌站、维修车间、仓库和铁件车间都已包括在内。

Mr. Armstrong: Well, it's detailed enough. According to the Method Statement you submitted, the precast concrete piles will be fabricated on site.

白先生：是的。2 台搬运混凝土桩的龙门吊会放在这。

Mr. Armstrong: What is the capacity of these cranes?

白先生：每台 10 吨。

Mr. Armstrong: It seems that if you use 8000 m^2 for precast yard as shown in this layout, there will be not enough area for other facilities.

白先生：这块场地只是短时间内用作预制构件厂。当桩基进行 3 个月后，我们会缩减预制构件厂的面积，将它变为木工房、仓库、维修车间和铁件车间。我们将在施工期间修建一个钢筋场，其他设施以后再建。

Mr. Armstrong: Is a bending shop included in your rebar yard?

白先生：是的。在钢筋车间中我们有钢筋切断机和钢筋折弯机，能加工的钢筋的最大直径为 40 毫米。

Mr. Armstrong: Fine. How many concrete batching plants shall be erected on site?

白先生：2 个。1 个功率为每小时 40 立方米，另一个为每小时 25 立方米。

Mr. Armstrong: Will the waste water from the concrete batching plant drain directly into the city drainage system?

白先生：当然不是。在进入市政排水系统前废水会通过集泥井过滤。

Mr. Armstrong: I cannot find where your site lab is located. It should not be omitted from your layout because it is an essential facility for quality control.

白先生：当然不会。你看，距搅拌站不远的地方就是试验室。试验室内将配备土工试验、混凝土试验和钢筋试验所需的全部仪器。

Mr. Armstrong: Where is site office?

白先生：在这儿。我们的办公室能容纳 40 人一起办公，业主办公室能容纳 15 人。

Mr. Armstrong: Is there enough space for a meeting room?

白先生：噢，真对不起。我们忽视这一点。如果在您的现场办公室旁边增加 40 平方米的会议室，您会不会介意？

Mr. Armstrong: Of course not. It's more convenient for us to communicate with each other.

白先生：我们会立即修改现场布置图，下周再提交给您。

Mr. Armstrong: Fine.

Dialogue 2：施工人员招录（音频：5.3.2）

Word Tips

招聘	recruitment
传染病控制	infectious disease control
稀罕的事物	white crow
高血压	hypertension
身体残疾	physical disability
艾滋病	AIDS
霍乱	cholera
肝炎	hepatitis
外立面	facade
蚊帐	mosquito netting

隔热	thermal insulation
疟疾	malaria
疟疾高发区	high-risk area of malaria
防滑瓷砖	anti-slip ceramic tile
太阳能热水器	solar water heater

【场景：方先生——项目公司副总经理，Jackson 先生——省劳动就业发展厅厅长，Morris 夫人——省卫生厅副厅长】

方先生：欢迎 Jackson 先生和 Morris 夫人。非常感谢你们在本地员工招聘的第一天光临我们的办公室。

Mr. Jackson: This is a big day of your company and of my department either. Thousands of young people in this province have been looking for employment for long time. How could I be absent?

Mrs. Morris: Me too. I am very delighted that hundreds of young people will find their jobs here, but in the meantime, I am very concerned about the tremendous challenge of infectious disease control.

方先生：Jackson 先生，我希望你能帮助解决我们在当地工人招聘中遇到的问题。一些当地的应聘者甚至连身份证都没有。

Mr. Jackson: Oh, this is not a white crow in this country. ID cards have not covered some remote rural areas yet. Don't worry, Mr. Fang, I will call the police chief to send a sergeant here to issue ID cards to them.

方先生：您帮了我一个大忙啦。谢谢您。

Mr. Jackson: Up to now, how many local workers have been recruited?

方先生：这次我们公司打算招聘 500 名本地工人。到目前为止，我们已经收到了 700 多份求职申请。招聘程序非常严格，包括申请、身份认证、体检和面试几个步骤。我们只有等到该申请人通过所有的步骤才能确认是否录用他。因为我们三天后才会有体检报告，所以这会儿我还不能确定已经录用的当地工人的人数。

Mr. Jackson: Three days? It takes quite long time. What kind of health examinations are the local applicants undergoing?

方先生：除了一般的体检，我们主要要求做两套体检项目。第一套用于确定应聘者是否患有心脏病、高血压、身体残疾等不适合在建筑工地工作的疾病；第二套用于排除艾滋病、霍乱、肝炎等传染病的患者。健康检查方案是根据省卫生厅的建议制定的。非常感谢 Morris 夫人。

Mrs. Morris: That is my duty and pleasure. Since you are going to mass hundreds of workers in your camp, I hope to inspect the accommodation for the local workers with Mr. Jackson together.

Mr. Jackson: Good idea. Mr. Fang, let's go to your camp for the local workers right now.

方先生：很高兴接受您的检查。那再好不过了，前面就是本地工人的营地，我们上个月才建的。

Mr. Jackson: Aha, the houses are all built in containers, aren't they?

方先生：是的，这个营地共用了 150 个集装箱，其中宿舍用了 125 个，食堂、淋浴室、洗衣房、厕所用了 25 个。

Mrs. Morris: Very impressive. You've painted the facade with lovely colors and planted trees in each yard. I'd like to see the fitments inside the rooms.

方先生：这是由一个集装箱改造而成的标准宿舍。我们切割出了门窗洞口，用隔热材料对内墙做了装修，并为宿舍安装了木地板和空调。

Mrs. Morris: There are four beds, one table and one shelf. Why I couldn't find mosquito netting here? This is indeed a high-risk area of malaria.

方先生：我们已经注意到了，并且已经购买了足够的蚊帐。每一个本地工人一旦被正式录用就会有一套的。

Mrs. Morris: Let's have an inspection to the shower rooms and toilets then.

方先生：我们建了 10 个淋浴房，每个淋浴房有 5 个淋浴喷头。你看，淋浴房的屋顶上有三个太阳能热水器。淋浴房的地面是防滑瓷砖，实际上，就连食堂和厕所也都一样。

Mr. Jackson: You've built a good camp for local workers. I will come again three days later when you completed your local workers recruitment.

第四节　篇章口译

Passage 1：如何进行施工准备（音频：5.4.1）
Word Tips

raw	生的，未加工的
preliminary work	初步工程，准工程
scrub	矮树
jungle	林，密林
uproot	根，迁离
orientation	方向，定向
trench	沟，沟渠
lay out	展示，安排
sediment	沉淀物
grubbing	除根，挖除
storm drain storm sewer	雨水道，雨水沟
sewer pipe	污水管，下水管
soil erosion	土壤侵蚀，水土流失
shovel	铲除，用铲挖
detention	拘留，延迟
basin	盆地，盆，水池
side slope	边坡，坡
dozer	推土机
debris	碎片，残骸
elevation	海，高地
disturbance	干扰，骚乱
manhole	人孔，检修孔
primer	原始物
polyurethane foam	聚氨酯泡沫，海绵
coating	层，包衣

aerosol	喷雾器,气雾剂
vapor	蒸汽
interior	内部的,国内的
sheeting	薄片,挡板
harden	变硬,变坚固
elapse	过,消
pressure	压强
installation	安装,装置
vacate	空出,腾出

Site preparation is defined as preparing for necessary operations that are required to develop buildings, parking lots, roads, etc. on raw land. Site preparation is one of the preliminary works to be done for starting constructional works. The selected construction site for building needs to be prepared properly.

The following factors are considered for the construction site preparation:

1. The first step of site preparation is to remove all the scrubs or jungle if there exists any on the site for building construction.

2. The whole area will be roughly leveled.

3. The holes of the construction site will be filled with sands or rammed earth and leveled off as required or redirected by the authority.

4. As a part of site preparation, the trees will be cut off and their roots are totally uprooted as directed by the authority.

5. Before starting the work, permanent bench marks must be established at a suitable point in the construction site.

6. The orientation and trench lines of the building should be correctly laid out in the construction site and the location for the storage and stacking of the materials should be definitely set on the ground in the site.

Preparation includes several different tasks like sediment control, clearing and grubbing, storm drains, water and sewer pipes, soil erosion management, rock removal, weed removal, underground utility and other tasks. It is very important to protect the quality of water, and control soil erosion and sediment once the location is selected. Many locations require storm permitting, and all erosion control measures must be placed and inspected firsthand before cutting the first tree and shoveling off the dirt from the site.

Storm water management system is very complex because the detention basins are complex in nature and because it has side slopes and flat grade bottom. Clearing is cleaning the pathway in the site so others can pass easily. Usually, the limit of clearing is done with the help of GPS dozer. Burning the debris is the traditional method but this practice is fading away. Moreover, in most areas, the air pollution standards will stop from burning the debris.

Site preparation requires lots of planning and using of heavy machinery. You never know when you will need one item and tomorrow you need something else. So preparing a plan in advance of the project is vital. Most common tasks in excavation are clearing lots for houses, laying pipes, fixing water leaks, digging foundation and others which require a teamwork.

When laying a pipe, it requires skill and a good plan. After you dig the trench, you have to make sure that the angle of elevation is correct. Pipes specification must match the blueprint. Several different pipes which are used for water, sewer and storm drains need to be placed carefully.

In order to start the job, you must get a permit. You can only start the work once you get the permit to work on the site. In some cases, you have to mention how much disturbance the site preparation will create. For that you will have to get an environmental study document to start the work according to your plan. When you are making excavation on the site, you might dig and remove land to create a pond. This also calls for skilled and experienced workers. Manholes are opening holes or an access points for performing maintenance and created for purposes like storm drains, sewers, telephones, electricity.

HVAC systems are typically shut down during some parts of roof preparation, as well as during application of primers, spray polyurethane foam, and coatings. System shutdown stops the drawing of dusts, aerosols and/or vapors into interior spaces. Once the HVAC system is shutdown, seal the air intakes with plastic sheeting and tape, which will prevent dust and spray from entering the intakes. Keep the plastic sheeting in place at least several hours after the spray application is completed, typically 24 hours or more; a longer period may be appropriate for coatings, depending on when the coatings have hardened or set and when they no longer emit vapors. The HVAC system should not be restarted until appropriate time has elapsed and the plastic sheeting and tape are removed.

There are many factors to consider when planning a SPF high pressure installation. Will the work take place in an occupied building or a building under construction? Will the building be vacated? Will other trades workers be present at the time of application? Will the application take place indoors or outdoors? What is the size of the work area—a large open area or an attic or crawlspace with limited ventilation?

In conclusion, when you start working on the site you need to take the site itself rules and regulations, first aid and team spirit into account. It could be fun work if you can easily handle the heavy machinery and work as a team.

Passage 2:施工规划过程的三个阶段（音频:5.4.2）

尽管任何给定的项目都有许多可能的计划,但制定一个好的施工计划是非常具有挑战性的。虽然过去的经验是建筑规划的良好指南,但每个项目都可能有特殊的问题或机会,可能需要相当大的独创性和创造力。此外,施工过程是动态的,这意味着施工计划必须随着施工的进行而修改。

通常,施工规划过程包括三个阶段:估算阶段、监控阶段和评估阶段。它开始于规划师开始计划项目的施工时,结束于施工过程的结果评估完成时。

估算阶段包括实施施工成本和工期估算。在此阶段,必要活动的资源需求也将由规划师进行估算,规划师将对现场特征带来的不同条件进行仔细彻底的分析,以确定最佳估算。

在施工规划过程的监控阶段,施工经理必须不断跟踪正在进行的活动的持续时间和成本。有时施工进度可能推迟或提前,但成本不在估算范围内或低于估算。因此,在施工完成之前,必须不断进行监督。当施工过程完成并将相关信息提供给规划者时,施工规划过程的第三阶段就可以开始了。

评估阶段是根据估算对施工过程的结果进行评估的阶段。规划者一定要意识到并处理估算阶段的不确定性。只有当施工过程的结果已知时,才能正确评估估算的有效性。此外,在施工规划过程的最后阶段,规划者可以确定估算是否正确。否则,他应该在未来的施工规划中做出调整。

第五节　口译技能——记忆与预测

一、训练目标

本单元训练学生根据发言体裁的惯用结构积极预测,提高记忆效果。

二、原理及技巧讲解

人类储存信息的记忆种类有两种:短时记忆和长时记忆。

短时记忆,顾名思义,只能在脑海中留存很短的时间。举个例子,去快递柜取快递,取件码就储存在短时记忆中,留存时间也就是取件输入时的那几秒钟。除了有限的留存时间,短时记忆的另一个特点是其容量有限。心理学研究表明,普通人的短时记忆容量只有"7±2"个单位。比如,取件码分为6位数和8位数两种,对于后者,很多人需要分两次记忆和输入,特别是在不太专注的情况下。

长时记忆指信息、知识、经验、操作能力和价值观等在我们的大脑中永久储存,其容量也是无限的。在接下来的练习中,我们谈到"已有知识""知识储备""背景知识""常识""经验""个人体验"等,指的都是译员长时记忆的具体方面。

我们常说一个译员要有百科全书般的知识储备,读万卷书、行万里路,就是因为人类拥有大量共性的知识和经验。尽管译员会接触到各种各样的主题,但没有任何发言是全新的,共有的背景知识和经验是口译理解和记忆的基础。因此,在听辨过程中,要善于在接收的新信息与已有知识和经验间建立联系。联系的前提是听辨的主动性,在听辨中寻找信息间的相关性,并且激活相应的背景知识。以数字为例,如果阅读的时候没有积极思考,可能会忽略数字间的联系。

记忆组块的训练方式是复述。与概述不同,复述时除了要包括发言人的主旨和主要信息点,学习者还需要纳入尽量多的细节。

预测也是口译中常用的策略。首先,我们可以通过常见搭配,做词汇句法层面的预测;其次,我们能通过语境信息,如发言人、场合和主题,去预测发言内容。更重要的是,口译中有相当一部分发言是程序性或礼仪性的,有固有特点,更便于预测。学者 Gillies 对此深有体会:

Many speeches that you will have to interpret consecutively will be ceremonial... speakers, even improvising, will often stick to certain conventions. Some speakers a naturals, some have speeches written for them, others learn how to speak from guide books and manuals (and some, who perhaps should, unfortunately don't!)

Gillies 所说的 "conventions" 呼应了一个语篇分析中常见的概念:体裁(genre)。我们来看两位学者对 "genre" 的定义:

Genre is a socially ratified way of using language in connection with a particular type of social activity.

Genre is a purposeful cultural event that is realized through schematic structure and realization pattern.

三、示例精讲

我们选择了一场投资推介会的开幕致辞作为练习材料。在听录音前,请大家首先从发言目的、发言人的身份预测发言内容。

在课堂上讨论后,学生认为该材料可能包括以下四个部分:

(1)欢迎/致谢参会者;

(2)介绍本次会议;

(3)邀请参会者前往投资;

(4)再次感谢参会者。

你的预测与此相似吗?

我们来听第一篇材料:时任印度驻华大使苏杰生在 2011 年印度投资机遇推介会上的致辞。请分析发言结构并验证自己的预测是否准确。

发言原文:(音频:5.5)

I am very pleased to join you this afternoon at this Workshop on "Investment Opportunities in India".

We are all aware that India-China economic cooperation has grown dramatically in the last decade. Trade itself expanded around ＄3 billion in 2000 to ＄62 billion in 2010. Companies from each country are exploring the advantage of manufacturing in the other. The service sector has also shown greater awareness of business opportunities. We are looking at the prospect of expanding our banking linkages as well.

Against this backdrop, the role of investment is gaining importance as it represents the next stage of our economic engagement. There are four broad reasons why it is important for Chinese companies to invest in India today.

First, having initially developed the market through trade, they need a local presence to maintain and service their existing and future clientele. In fact, their reputation depends on how effectively they do so. Second, if they wish to increase their commitment to India on a long-term basis, they can do so best in collaboration with Indian partners. In many areas, such localization may enhance their branding and will certainly assist in building a distribution network. Third, India can be a base from which Chinese companies can export beyond into the wider region, particularly the Middle East, Indian Ocean region and Africa, and fourth, Chinese companies with excess capacities facing rising costs and falling demand at home would find India's deficits, human resources and cost consciousness the right mix to set up business overseas.

You will hear from speakers today the prospects and possibilities for such investments. We have also organized experience sharing sessions that would be useful. I am sure you will have a productive Workshop.

Thank you again for joining this conference, wish you a very fruitful afternoon.

这篇发言包括以下几个部分：

(1)欢迎参会者；

(2)主办国与中国的紧密关系(投资基础)；

(3)主办国作为投资目的地的优势(四个点)；

(4)介绍接下来的议程；

(5)表示感谢。

可以看出,学生的预测是基本准确的,着墨最多的第(2)、第(3)部分,目的的确就是向投资者发出邀请,但直接邀请并不符合发言人的身份,这也是预测时需要考虑的因素。

第六节　课后练习

（音频:5.6）

Ⅰ. Technical Terms Interpretation

1. 施工流水作业法

2. 施工面积

3. 项目预算编制

4. 项目支助业务

5. 施工平面图

6. 施工期

7. 施工企业

8. 施工前阶段

9. 施工区

10. 施工缺陷

11. working drawing estimate

12. construction documents

13. construction error

14. fabricating yard

15. detail of construction

16. project preparation facility（PPF）

17. project loan

18. menu

19. project formulation

20. total tender value of project

Ⅱ. Sentences Interpretation

1. 我们应该按照工程项目的总进度表（建设进度表）工作。

2. 这个合同的有效期将从 2027 年 12 月 30 日开始。

3. 卖方将于五月份向买方提供初步（最终）技术文件。

4. 我们主要的计划工作项目包括费用预算和施工进度。

5. 下个月我们将召开设计数据收集(初步设计、最终设计)会议。

6. 现场安装工作(土建工作)将自今年十月开始至明年六月一日完工。

7. 这座工厂的交工验收日将在 2028 年 4 月 6 日。

8. 卖方操作组(专业工作组)将在现场,一直工作到生产符合保证条件。

9. 我们必须使工厂通过试运转并最终投入工业生产。

10. 因缺少材料(施工机械、安装工具),我们只能改变计划。

Ⅲ. Dialogue Interpretation

张天:你好,徐辉。你这几天过得怎么样? 对不起,我没有太多时间和你说话。

Xu Hui: I'm OK. What are you busy with?

张天:我最近忙着找项目。

Xu Hui: Is there any progress? You know, I'm new here and I want to get into a project as soon as possible.

张天:没有什么进展。但这很正常。

Xu Hui: Oh, I thought there would be a lot of projects here as Liberia is a developing country and is developing at a high speed.

张天:你说得对,所以我们有很多机会。在此之前,我认为你可以先尝试获得更多与项目相关的知识。

Xu Hui: That's what I'm thinking about, and I just want to ask you what we should do before a project starts.

张天:嗯,任何项目都有一个施工前阶段,包括规划、预算和获得许可。

Xu Hui: Yeah, I know something about them. Planning is such a complicated process that it often takes years, especially when public funds are involved.

张天:是的。你似乎对一个项目很了解。

Xu Hui: No, I'm a closet strategist and I have never actually touched any project. So, more site work is needed.

张天:不要太谦虚了。我今天下午有空。你为什么不和我一起去皇家酒店喝杯咖啡?

Xu Hui: I would like to, and may we continue the topic on project?

张天:当然。

Ⅳ. Passage Interpretation
施工前准备工作的主要部分

施工前准备通常包括技术准备、材料准备、劳动力准备、施工现场准备等。每一个部分都将详细说明如下。

一、技术准备

它是施工前准备工作的核心。任何技术错误都可能导致严重事故,造成巨大的生命、财产和经济损失。因此,技术准备工作必须一丝不苟。一般来说,技术准备包括审查和熟悉施工图纸和相关设计资料;调查和分析施工现场的原始数据;建立数据管理系统、测量控制系统和质量检验系统等。

二、材料准备

材料是工程顺利施工的物质基础。建筑材料、建筑安装设备、生产加工设备等材料的准备工作以及零部件的加工准备工作必须在项目开工前完成。在准备材料时,应根据不同材料的不同要求,确定来源,妥善规划运输和安排储存,以满足施工需要。

三、劳动力准备

从范围上讲,有整个建筑企业的劳动力准备,也有大型综合性建筑项目的劳动力准备和小型、简单的单位工程的劳动力准备。从内容上讲,劳动力准备包括成立拟建项目的领导机构;组织一支精干的施工队伍;向施工队明确相关的设计、规划和施工技术;建立和完善行政管理体制。

四、施工现场准备

建筑工地是所有参与者有节奏、平衡、连续地工作的空间,以实现高质量、高速度、低消耗的目标。施工场地的准备主要是为拟建项目创造和保证有利的施工和物质条件,包括在施工现场做好控制网测量;建设水电等基础设施;修路和平整地面;对施工现场进行补充勘探;建立临时设施;安装和测试机器和工具等。

第六章　工程材料采购与设备租赁

工程材料和设备的采购及租赁是海外工程总承包项目的重要组成部分,在投标前,承包人一般要在工程所在国及邻近国进行市场调查,寻找中标后工程所用材料设备的货源。中标后,根据施工组织设计编制的物资供应总计划,应交给物资部,物资部接到计划后着手实施材料设备的采购工作。在项目实施过程中,工程设备和材料的采购管理主要包括询价,制定采购计划,加强市场调研、合理选择供应商,材料价格的控制要点,材料的进场检验以及材料的现场管理,租赁须知,海关代理。同时,选取一名优秀的代理人对于加快施工进度、保证工程质量、降低工程成本、提高经济效益,具有十分重要的意义。

第一节　背景知识

一、询价

采购的第一步是询价,包括在承包人所属国、工程所在国及第三国进行询价。凡在工程所在国能采购到的材料设备,只要经济适用,且符合工程合同的规定,就应尽可能在当地询价并采购。询价的方式一般通过电邮或传真向几家供货商发出询价函,从中择优。询价时一定要将材料或设备的规格写清楚,以免购错,耽误工程的施工。对于大量的、重要的以及定期的采购,最好派专人谈判,以争取最好的供货条件。

二、编制采购计划

项目部依据项目合同、设计文件、项目管理实施规划和有关采购管理制度编制采购计划。采购计划包括:采购工作范围、内容及管理要求;采购信息,如产品或服务的数量、技术标准和质量要求;检验方式和标准;供应方资质审查要求;采购控制目标及措施。

三、加强市场调研,合理选择供应商

一是审核查验材料生产经营单位的各类生产经营手续是否完备齐全;二是

实地考察企业的生产规模、诚信观念、销售业绩、售后服务等情况;三是重点考察企业的质量控制体系是否具有国家及行业的产品质量认证,以及材料质量在同类产品中的层级;四是从建筑业界同行中获得更准确、更细致、更全面的信息;五是组织对采购报价进行有关技术和商务的综合评审,并制定选择、评审和重新评审的准则。

四、材料价格的控制要点

一是对材料的采购价格进行控制。企业应通过市场调研或者通过咨询机构,了解材料的市场价格,在保证质量的前提下,货比三家,选择较低的材料采购价格。二是对材料采购时的运费进行控制。要合理地组织运输,比较材料采购价格时要把运输费用考虑在内。在材料价格相同时,就近采购材料,选用最经济的运输方法,以降低运输成本。三是要合理地确定进货的批次和批量,还要考虑资金的时间价值,确定经济批量。对于一些特殊材料(如炸药),若以离岸价购买,则必须考虑到运输的安全问题。四是选择承运人一定要慎重,不要单单根据运输费的报价来确定。

五、材料的进场检验

建筑材料验收入库时必须向供应商索要国家规定的有关质量合格证明及生产许可证明。项目采用的设备、材料应经检验合格,并符合设计及相应的现行标准要求。材料检验单位必须具备相应的检测条件和能力,经省级以上质量技术监督部门或者其授权的部门考核合格后,方可承担检验工作。采购的产品在检验、运输、移交和保管等过程中,应符合职业健康安全和环境管理要求,避免对职业健康安全、环境造成影响。

六、材料的现场管理

材料存放管理。建筑材料应根据材料的不同性质存放于符合要求的专门材料库房,应避免潮湿、雨淋、防爆、防腐蚀。一个建筑工地所用材料较多,同一种材料有诸多规格(比如:钢材从直径几毫米到几十毫米,有几十个品种;水泥有标号高低之分,品种不一;各种水电配件品种繁多),所以各种材料应标识清楚,分类存放。

材料发放管理。要建立限额领料制度,对于材料的发放,不论是项经部、分公司还是项目部仓库物资的发放,都要实行"先进先出,推陈储新"的原则,项目部的物资耗用应结合分部、分项工程的核算,严格实行限额领料制度,在施工前必须由项目施工人员开签限额领料单,限额领料单必须按栏目要求填写,不可缺项。对贵重和用量较大的物品,可以根据使用情况,凭领料小票分多次发放。对于易破损的物品,材料员在发放时需详细查验,并由领用双方在凭证上签字认可。

组织管理。施工中的组织管理是现场材料管理和管理目标的实施阶段,其主要内容如下:

(1)现场材料平面布置规划,做好场地、仓库、道路等设施的准备;

(2)履行供应合同,保证施工需要,合理安排材料进场,对现场材料进行验收;

(3)掌握施工进度变化,及时调整材料配套供应计划;

(4)加强现场物资保管,减少损失和浪费,防止物资丢失;

(5)施工收尾阶段,组织多余料具退库,做好废旧物资的回收和利用。

七、租赁须知

施工设备是关乎施工企业是否能够正常施工作业的关键。在现代社会的生产活动中,技术与设备的更新换代关系着企业在未来很长一段时间的盈利情况。但是现代社会中技术的更新速度远远超过施工企业的更新速度,往往会出现施工企业刚采购的新设备很快被其他设备取代的情况。因此租赁也是施工设备的供应渠道之一。施工设备的租赁一般用于两种情况:一种是在工期前期,在所购设备还没有到达现场之前,为了满足紧急开工的需要而租赁;另一种情况是,有些设备虽然是施工所必需的,但使用时间短,利用率低,对于此类设备也常常采用租赁方式。租赁施工设备除了要注意租赁费,还一定要派机械工程师去检查设备的运行情况,以确保租赁设备在现场的工作效率和出勤率。设备和材料的租赁有优点,也有缺点。优点是不会对施工企业的现金流造成影响,不用害怕技术革新,不需要额外招聘相关设备维修和保养人员。缺点就是风险、压力大:过于依赖租赁市场,有时会存在经济压力,需要向租赁公司支付设备租赁和技术人员资金;遇到特殊情况,工期受到影响,企业面临资金困难。

八、海关代理

在国外承包工程,往往需进口大量物资,因而通常会雇用一个海关代理,帮助承包人办理清关手续。选择代理人一定要看其声誉和能力,以保证货物到港后能很快转运到工程现场,并及时投入使用。关于代理费,声誉好的代理人

（行）一般都有自己的收费标准。由于工程承包需进口大量物资，每批货的价值也比较大，通常对代理很有吸引力，因此，双方可在商定收费标准后，再签订代理协议书。

第二节　相关词语与表达

序号	中文	英文
1	钢丝网门	chain link
2	两面包铁皮门	door clad with sheet iron on both sides
3	保温门	thermal insulation door
4	保密性	confidentiality
5	保证金	advance payment
6	报价	quotation
7	备选方案投标	alternative bid
8	标准招标文件	Standard Bidding Documents(SBDs)
9	采购	procurement
10	采购代理	procurement agent
11	采购公告	procurement notice
12	采购计划	procurement plan
13	采购决定	procurement decision
14	承包商	contractor
15	初步描述	preliminary description
16	初步设计	preliminary design
17	履约保函	performance security
18	履约证书	performance certificate
19	投标函	letter of acceptance
20	进度报告	progress report
21	房地产开发商	real estate developer
22	违约罚金	liquidated damages
23	预付款	advance payment

续表

序号	中文	英文
24	有权,有资格	be entitled to
25	工程量清单	bill of quantities
26	符合	correspond with
27	到期债款	debt due
28	应付债款	debt payable
29	缺陷责任期	defects liability period
30	暂扣款	retention money
31	单价表	schedule of rates
32	投标保证金	tender security
33	偏离合同规定	deviation form
34	成本补偿合同	cost-reimbursement contract
35	带纱扇窗	window with screen sash
36	固定百叶窗	louver window
37	遮阳式窗	awning window
38	卷帘百叶窗	rolling shutter
39	挡风雨条	weather strip
40	窗铁栅	window guard
41	窗插销	sash bolt
42	可拆铰链	loose joint
43	窗开关调节器	window adjuster
44	地弹簧	floor spring
45	门弹弓	door closer-spring
46	推拉门滑轨	sliding door rail
47	单元门锁	unit lock
48	磁卡门锁	card lock
49	推拉把手	push and pull brace
50	门推板	push plate
51	踢板	kick plate
52	定门器	door stop
53	闭门器	door closer

续表

序号	中文	英文
54	弹子锁	night latch
55	转轴	pivot
56	钥匙孔盖	key hole escutcheon
57	气窗联动开关	window gearing
58	手动开关器	hand opener
59	电动开关器	electric opener
60	箱形楼梯	box stair
61	悬臂式楼梯	bracketed stair
62	悬空楼梯	hanging stair
63	螺旋楼梯	spiral stair
64	盘旋楼梯	winding stair
65	交叉布置的自动楼梯	crisscross arrangement of moving stair
66	平行布置的自动楼梯	parallel arrangement of moving stair
67	回转式楼梯	turn stair
68	双分叉式楼梯	bifurcated type stair
69	狗腿式楼梯	dog-legged type staircase
70	带梯井式楼梯	open-well staircase
71	中柱螺旋式楼梯	helical type stair with newel
72	单跑式楼梯	staircase of straight flight type
73	防火梯	fire escape staircase
74	专用太平梯	special emergency staircase
75	金刚砂防滑条	non-slip emery insert
76	楼梯踏步防滑条	non-slip nosing
77	楼梯毯压棍	stair carpet rod
78	固定楼梯地毯夹	stair clip
79	板条顶棚	lathed ceiling
80	露梁顶棚	beam ceiling
81	拱顶式顶棚	vault ceiling
82	胶合板顶棚	plywood board ceiling
83	抹灰金属网顶棚	plastered metal-lath ceiling

续表

序号	中文	英文
84	平顶隔栅	ceiling joist
85	顶棚吊筋	ceiling hanger
86	铝合金压条	aluminium bar
87	顶棚通风孔	air drain of ceiling
88	顶棚的木压条	ceiling panel strip
89	伸缩缝盖条	expansion joint sealing
90	屋面材料	roofing materials
91	防潮材料	damp-proof materials
92	沥青防潮层	asphalt damp-proof course
93	聚乙烯薄膜	polythene film
94	聚氨酯薄膜	polyurethane film
95	隔热保温材料	heat-insulating materials
96	聚苯乙烯片	polystyrene sheet
97	玻璃纤维板	fiberglass sheet
98	绝热屋面板	insulating fiberboard
99	石棉水泥板	asbestos-cement sheet
100	膨胀软木片	expanded cork sheet
101	钢筋	rebar
102	保温箔片	thermal foil
103	密闭条	sealing rope
104	油毡覆盖面层	felt carpet
105	防水材料	waterproof material
106	不透水层	waterproof layer
107	不透水接头	water-tight joint
108	屋面油毡	roofing felt
109	油毛毡衬	sarking felt
110	浸渍毡	impregnated felt
111	油毡纸	felt paper
112	灌注石油沥青	poured asphalt
113	松节油灰泥	gum-spirit cement

续表

序号	中文	英文
114	防凝水内衬	anti-condensation lining
115	沥青衬里	bitumen-lining
116	沥青嵌缝料	bitumen-sealing compound
117	黏结剂	binding agent
118	沥青胶泥填缝	molten bitumen filler
119	冷乳化沥青	cold emulsified bitumen
120	焦油沥青涂层	tar membrane
121	乳化液涂层	emulsion membrane
122	环氧涂层	epoxy membrane
123	五层作法	five-ply
124	冷铺	cold application
125	热铺	hot application
126	脊顶	ridge crest
127	檐瓦	verge tile
128	滴水瓦	drip tile
129	披水板	cover flashing
130	天沟托板	layer board
131	封檐板	eaves plate
132	抹灰板	plastering plate
133	抹灰板条	lath and plaster
134	抹灰罩面层	setting coat
135	珍珠岩砂浆抹面	pearlier plaster finish
136	水砂粉饰	sand plaster
137	砂饰面	sand finish
138	喷珍珠岩饰面	pearlier spraying
139	水泥砂浆垫层	cement and sand cushion
140	拉毛	stucco
141	水刷石饰面	granitic plaster
142	石灰浆刷	lime brush
143	瓷砖铺面	pave with tile

续表

序号	中文	英文
144	贴釉面砖	facing glazed brick
145	贴陶瓷砖	facing ceramic tile
146	木材等级	timber grade
147	涂料成分	paint ingredients
148	防锈层	antirust coat
149	乙烯基树脂涂料涂层	vinyl-resin paint coating
150	防腐蚀涂料	anticorrosive paint
151	可赛银粉	calcimine
152	沥青冷底子油	asphaltic base oil
153	防滑漆	non-slip paint
154	氧化锌底漆	zinc oxide primer
155	室外涂饰	interior paint
156	雪花粗面涂饰	frosted rustic work
157	蜂窝状漫涂	combed stucco
158	斜遮雨板	raking flashing
159	遮阳花格	sun shade grille
160	网眼钢皮抹灰隔墙	expanded metal lath partition
161	石膏墙板	gypsum wall board
162	刺篱	thorn hedge
163	定制的产品	custom made
164	高于标准价格采购请求	premium price request（PPR）
165	报价请求,询盘	request for quote（RFQ）
166	供应商早期介入	early supplier involvement（ESI）
167	成本模型	cost modeling
168	标准价格	STD price
169	缺货成本	stock-out cost
170	指定供应商	awarded supplier
171	提交上级处理	escalate to higher level
172	手动下订单	cut hard order
173	客户需求提前或增加	customer demand pull-in

续表

序号	中文	英文
174	分销商	distributor
175	厂商	manufacturing
176	经纪商	broker
177	预估每年需求量	estimated annual usage（EAU）
178	停产	line down
179	商品不含税零售价	unit retail
180	商品状态	item status
181	毛利率	MU%
182	商品类型	item type
183	商品建立日期	create date
184	供应商号码	vendor number
185	供应商货品编号	vendor item number
186	供应商箱包装成本	VNPK COST
187	销售金额	POS Sales
188	自动采购系统	automated purchasing system（APS）
189	售后服务	after service
190	供应商	supplier
191	商展会	trade fair
192	托收	collection
193	委托人	principal
194	托收银行	remitting bank
195	付款交单	Documents against Payment（D/P）
196	远期付款交单	Documents against Payment after sight
197	接受订货	accept order
198	备选材料	alternative materials
199	年销售额	annual sales
200	出口许可证申请书	application for export permit
201	管制货物进口申请书	application for importation of controlled commodities
202	外货进口报单	application for import of foreign goods
203	汇票	bank draft

第三节　对话口译

Dialogue 1（音频:6.3.1）

Word Tips

retail price	零售价
wholesale price	批发价
deformed rebar（bar）	螺纹（变形）钢筋
call for tender（call for bid）	招标
BOT：Build-Operate-Transfer	建造—运营—转交
BOOT：Build-Operate-Own-Transfer	建造—运营—拥有—转交
deferred payment	延期付款
guarantee	担保书
government bulletin	政府公告
round rebar（bar）	光圆钢筋

【场景:In building material market, A is the purchasing manager, he wants to purchase some rebar as part of the project, B is the boss of the building materials market.

在建材市场,A 是采购部经理,想要采购一些钢筋作为项目零件,B 是建材市场的老板。】

A:Excuse me, sir. Does your shop supply rebar?

B:是的,我们提供各种直径和长度的钢筋。你需要哪一种,是光圆钢筋还是螺纹钢筋?

A:Round ones with 6 mm diameter, deformed ones with larger diameter. May I know the price of the rebar?

B:当然可以,您看这是现在的零售价目表。一般来说,光圆钢筋每吨 225 美元,螺纹钢筋每吨 230 美元。

A:What about wholesale prices?

B:批发的价格取决于您订购的数量,您定的物品越多,您获得的报价越便宜。您需要多少吨?

— 111 —

A: If the price is good, I will order at least 5,000 tons for our project.

B: 达到这样的数量,我给你光圆钢筋每吨 210 美元,螺纹钢筋每吨 215 美元的价格。

A: The transportation to the site is already included in the price, I believe.

B: 这很难答应你,请问你的工地在哪儿?

A: Fifteen kilometers away from here.

B: 这样的话,您必须自己承担运输费用。

A: Where can I buy crushed stone and sand?

B: 就在山下的采石场。

A: Thank you.

Dialogue 2（音频:6.3.2）

Word Tips

leasing and hiring business	设备租赁业务
road rollers	压路车
heavy construction equipment	重型施工设备
bulldozers	推土机
motor graders	电动平地机
dump trucks	自卸车
excavator's bucket	挖掘机铲斗
telegraphic transfer	电汇

【场景:In order to meet the requirements of the road construction plan, it is necessary to complete this paving work as soon as possible. Due to insufficient equipment on the construction site, the project manager decided to rent one excavator and two road rollers. Mr. Hu from the procurement department is inquiring with Mr. Liu, the manager of the construction equipment leasing company, about equipment leasing matters.

为了达到道路施工计划的要求,必须尽快完成这项铺路工作。由于工地设备不足,项目经理决定租用一台挖掘机和两辆压路车。采购部的胡先生(以下简称 H)正在向建筑设备租赁公司的经理刘先生(以下简称 L)询问设备租赁

事宜。】

H:刘先生有人告诉我,您负责这个公司的设备租赁业务。我来这儿看看是否能从你们这里租赁一些我们所需要的设备。

L:You've come to the right place. We have a large amount of imported heavy construction equipment and can rent them to you any time you like.

H:都有什么设备呀?

L:We have bulldozers, road rollers, excavators, motor graders, dump trucks, and loaders, and so on.

H:这次我考虑租用一台挖掘机和两辆压路车。你们的挖掘机铲斗的容量是多大?

L:The excavator is a 235 Caterpillar and has a bucket of about 2.8 cubic yards.

H:它们各自的租赁费是多少?

L:It depends on how you rent it and whether you want to rent it on a monthly basis, by the hour or by the square meter. If you rent it on a monthly basis, the excavator costs 30000 yuan per month and the roller costs 24000 yuan per month. Charged by the hour, the excavator costs 30 yuan per hour and the roller costs 25 yuan per hour. Charging by square meter is 5 yuan per square meter for an excavator and 3 yuan per square meter for a roller.

H:租费里包括燃料费和操作员的费用吗?

L:Yes, and maintenance as well.

H:哪一方负责将设备从你们库房运到我们现场呢?

L:If the hire term exceeds 280 working hours or one working month, we will transport the equipment by trailer at our own cost. Otherwise, we will charge 500 yuan for the transportation of each machine.

H:工作时间你们通常怎样记录?

L:Our operator fills in the time sheets which are countersigned by your site superintendent. We have set forms of time sheets. Every time sheet will be in duplicate, one copy is kept by each side. Payment will be made according to the time sheets.

H:谈到支付,你们要求多长时间支付一次? 以什么方式支付?

L: At the end of each week. We need to make some advance payment first, and we would like to receive payment by telegraphic transfer.

H:我们希望按月支付,因为我们从业主那里是每月得到支付。这次我们计划租用你们的设备三个月,这个时间可不算短。

L: All right. Do you mean you will hire more equipment in the future?

H:是的,并且很可能是长期租用,如果你们现在的租赁费降低 18% 的话。由于施工设计的变更,我们的开挖工作量大大增加了。我们正在考虑所需的附加设备是买还是租。

L: To show our friendship and goodwill to establish good relationships with you, I am willing to give a four percent discount of the rates I quoted you for the excavator and the loader. If you hire any equipment from us in the future for three months or more, we will reduce the rate by 15 percent.

H:很好! 下周一就把设备送到我们的工地,可以吗?

L: No problem!

第四节　篇章口译

Passage 1:材料采购和租赁会议（音频:6.4.1）

Hello everyone, welcome to today's meeting where we will discuss important matters regarding material procurement and leasing.

Firstly, let's talk about material procurement. Timely acquisition of high-quality materials is crucial for the success of the project. We need to ensure obtaining the required raw materials from reliable suppliers and delivering them within the planned time frame. To reduce costs and enhance efficiency, we plan to establish strategic partnerships with several high-quality suppliers.

Secondly, regarding leasing. Sometimes, we may need to temporarily lease e-quipment or facilities to meet the project's needs. When choosing leasing arrange-ments, we will focus on the suitability of the equipment, lease duration, and costs.

Through leasing services, we aim to flexibly adapt to the changing needs of the project.

Before making decisions, we will conduct thorough market research to ensure that the selected suppliers and leasing arrangements meet the project's requirements and can provide excellent service.

Thank you for your attention, and let's work together to ensure the success of the project!

Passage 2:采购步骤（音频:6.4.2）

Word Tips

一式四联	quadruplicate
采购申请单	purchase requisition
存档核实	archive verification
报价	quotation

采购工作流程中要规范的事项:

1. 所有的采购申请必须填写一式四联,采购申请单经部门经理核签后,整份共四联交给资产会计,资产会计复核后送董事。

2. 采购申请单一共四联,经审批批准后,第一联作仓库收货用;第二联采购部存档并组织采购;第三联财务部成本会计存档核实;第四联部门存档。

3. 审核采购申请单:收到采购申请单后,采购部应进行复查,以防错漏。

（1）核对签字。检查采购申请单是否由部门经理签字,核对其是否正确。

（2）核对数量。复查存仓数量及每月消耗量,核对采购申请单上的数量是否正确。

4. 邀请供应商报价。

Passage 3:周转材料须知（音频:6.4.3）

Word Tips

turnover materials	周转材料
thermal foil	保温箔片
heat-insulating materials	隔热保温材料

window guard	窗铁栅
sash bolt	窗插销
window adjuster	窗开关调节器
vinyl-resin paint coating	乙烯基树脂涂料涂层
roofing materials	屋面材料
damp-proof materials	防潮材料
discrepancies	差异

Rental management of turnover materials

1. The company implements two levels of management and one level of accounting for turnover materials. The Turnover Material Leasing Department is the supplier of the company's internal leasing market for turnover materials, responsible for the procurement, leasing, receiving and dispatching, storage, maintenance, accounting and other work of turnover materials. The project department of the company shall not purchase turnover materials without authorization. The turnover materials required for project engineering must be rented from the turnover material leasing department and shall not be directly rented out. The turnover material leasing department shall uniformly adjust the internal and external aspects to meet the construction needs of the project.

2. The turnover materials for centralized management include:

(1) Thermal foil;

(2) Heat-insulating materials;

(3) Window guard, sash bolt, window adjuster;

(4) Vinyl-resin paint coating;

(5) Roofing materials, damp-proof materials, etc.

3. Each project department should prepare a one-time material preparation plan for the newly opened project according to the construction organization design (or construction plan). Before the turnover materials enter the site, they should declare to the leasing department half a month in advance and sign a leasing contract with the turnover materials leasing department in accordance with the Contract Law and relevant construction management regulations, clarifying the responsibilities, rights,

business and rental fees, maintenance fees, compensation fees, transportation fees, and other charging standards of both parties, as well as the acceptance method for material entry and exit. The leasing department should prepare for supply in advance according to the contract provisions.

4. The project department should prepare a weekly entry and exit plan for turnover materials on a monthly basis based on the construction progress (mainly including usage time, quantity, supporting specifications, etc.), and submit it to the leasing department after being signed by the project leader. All plans and actual usage should be basically consistent. Any economic losses caused by discrepancies between the plans and actual usage shall be fully borne by the project department.

5. Both the project department and the turnover materials leasing department should establish a turnover material ledger, ensuring accurate names, specifications, entry and exit dates, quantities, and finally conducting acceptance and settlement.

第五节　口译技能——口译笔记技巧

科学的严肃性要求科技翻译严谨、准确。无论同声传译还是即席传译都要求译员在最短的时间内充分利用自己的语言优势,运用一定的专业知识将听到的信息快速处理,力求准确地表达出来。然而,纵然译员有较强的记忆力和敏捷的反应力,在长句、数字和长篇大段需要传译时仍然必须借助笔记以保证传递的信息准确无误。虽然大多数译员都知道口译中做笔记的重要性,但在实践中还是经常能看到一些初涉译海的新手在听完难句后未能记住要点及时传译,甚至出现主讲人说完了,自己却愣在那里的情况——笔者就曾有过这样的经历。究其原因,关键是没有用好手中之笔。口译过程中,为了减少漏译和误译,可充分借助笔记将说话者的意思较完整、准确地表达出来,提高口译质量。笔记之妙用在于:

一、记下数字,精确表达,避免误译

口译允许有一定的随意性和不精确性,但对数字的准确度要求极严,而英汉数字进位的不同给译员传译带来了一定的困难。因此,口译员们不妨借助手中的笔将数字记下,找出规律(如英文中 ten thousand 是"一万",one hundred million 表示"一亿",one billion 为"十亿"),准确传译。出现再大的数字,也可根据笔记按上述规律对号入座,即使一时难以表达,通过纸条或黑板写给对方看,甚至逐一读出也比误译强。如遇到"9846.35 亿元、7635.23 亿元"这两个数,可先将数字记下来,再按规律将小数点往前移一位,用 billion 表示,即984.635 billion、763.523 billion。当几个数字同时出现时,就更应该逐一记录下来,否则,记忆力再强也难以全部记住并准确无误地进行传译。如:"232.42 亿吨煤炭储量,其中褐煤储量 154.52 亿吨,无烟煤 40.32 亿吨,烟煤 35.58 亿吨。"仅这四个数字,若不依靠笔记很难全部准确译出。在传译过程中可采用边听边记的方法,待主讲人说完一串数字后,为了力求准确无误,应将自己记下的数字与主讲人核对一遍后再译。

另外,由于我国有的行业尚未完全使用国际标准单位,汉英单位的不一致也给数字的翻译增加了难度。如"××省水资源十分丰富,理论蕴藏量约 1.03 亿千瓦,占全国的 15.3%,可开发总装机容量为 9000 万千瓦,年发电量 4500 亿千瓦时"。即使记下了句中的四个数字,传译时也得费番功夫,将单位统一后再译,即按"×万千瓦"即"×MW"(如"10 万千瓦"为 100MW)的规律将"亿千瓦、万千瓦"转换成 GW 或 MW。遇到此类问题时一定要慎重,宁可慢一点,记下来数一数有几个零,再按规律逐步译出,以保证数字和单位的正确。对于焦耳(joule)、卡路里(calorie)、盎司(ounce)、加仑(gallon)、英里(mile)、英尺(foot)、亩(mu)、公顷(hectare)等,则可照译,把单位的换算留给交流双方处理,以免弄巧成拙。如"该公司占地面积 198 亩"可直接将"198 亩"译为"198mu",最多再补充"one mu is 0.667 hectares"。

在科技翻译中量化的概念有很多,口译时数字的翻译决不能马虎。译员必须从稍纵即逝的信息中快速捕捉,用笔将数字记下,克服数字表达的困难,按规律准确传译,否则,很可能译出了句子的意思却把关键数字忘掉了,稍一疏忽便会酿成大错,所以一定要慎之又慎。

二、记下关键词,简化长句,准确达意

口译中常有长句出现,若不借助笔记很难较完整地传递说话者的意思。须根据英语强调形合、汉语强调意合的特点,记下关键词,将长句简化,再适当增减词语,译出中心意思。如将汉语长句"分析受水库淹没影响的植物群落及植物物种,重点是受国家保护的珍稀植物和具有特殊经济、科研价值的物种"译为英语,由于口译的时间要求,只要在听的过程中快速记下句中的关键词"分析、植物群落、物种、重点、珍稀植物、物种",利用适当的句型将记下的关键词连接起来,或用适当的过渡词"and""but""which"等将长句分成两句,把听到的内容在关键词的基础上丰富完善,这样就可以将说话的意思基本表达出来了。全句可以试译为"Analysis shall be carried out for the plant community and species affected by the reservoir inundation, and emphasis shall be placed on the rare and precious plants protected by the State and species with special economic and research value."。也许这并不是最佳的表达,但我想外宾在特定的氛围下基本能明白其意义,而这也就达到了口译的目的。

英语长句汉译时也要依靠笔记才能尽量译得全面。有时外宾说话速度太快,不记下来很难译全,常常只能译出前半句或后半句甚至片言只语,使交流双方不知其意。对于译者而言,英语长句的汉译似乎比汉译英更难一些,因为译员的听力必须过关才能胜任(且不说存在听不懂外宾方言、俚语的困难)。译员为了将听到的关键词快速记下,有时可采用只有自己看得懂的符号或关键词的首字母甚至英汉混合记录。如"They desire to eventually establish a Co-ordination Center that provides minute-to-minute monitoring and co-ordination of the Power Pool operations, and the duties and functions of the Co-ordination Center shall be carried out by the Operating Committee until such time that the Board of Directors deems it necessary to establish a Co-ordination Center."，口译时笔者根据英语重形合、汉语重意合的特点,在理解的基础上将原句拆开,分成几个短句表述,并利用口语表达的随意性将长句简化,"长话短说",试将上句译作:"他们希望最终成立一个联网协调中心,对电网运行随时进行监测和协调。先由运行委员会负责实施该中心的任务和功能,直到董事会认为有必要成立协调中心时为止。"这种译法一样能达到口译效果。若没有笔记的帮助,很难将类似的句子译全。

三、抓住中心,记下大意,理清层次,避免漏译

即席传译时,有的主讲人很体谅译员,尽量讲得简短扼要便于翻译;但也有的主讲人对译员的水平估计过高,讲上几分钟,而译者又不便打断,只能借助手中的笔,迅速记下主讲人的讲话大意,理清讲话层次,抓住中心,采用直译、意译相结合的翻译技巧,用通俗易懂的语言逐层译出,避免漏译。

一次笔者随团就某项目的合作出国考察。考察结束时,代表团团长在外方的答谢宴会上发言,他一口气讲了近 5 分钟,基本上也就是整个考察过程的总结。从组团出发前外方对此次考察行程的安排征求意见、几易其稿,到到达后受到公司有关部门的盛情款待,以及考察期间有××先生的全程陪同和及时介绍,使我们深入细致地了解了公司的经营管理和技术力量,进而联想到双方未来的合作设想,最后又谈到将把此次考察结果向有关部门汇报。笔者边听边记,待主讲人说完后迅速将此段话归纳成以下几个层次:1.行程安排;2.日常接待;3.考察内容;4.合作设想;5.结语、致谢。传译时运用了口译的随意性和不精确性,未完全按照主讲人的讲话顺序,而是围绕上述 5 个层次将各层意思逐一译出,虽然没有逐字逐句翻译,但已将中心大意译出。

再如"Further to our meeting at the Second EPF in Vientiane on 12-23 December last year, I have consulted with officials from EGAT in Bangkok and ADB in Manila on the Study of Transmission Inter-connection between ××× and ××× via the third country. Both parties agreed in principle with my proposed arrangements which are outlined in the enclosed copy of the letter to the Governor of EGAT dated 3 January. EGAT has agreed to participate in the feasibility study of the transmission inter-connection between ××× and ××× which will be financed by ××× commercial bank."这一段话,笔者根据记录的要点"电力论坛会、输电线路经第三国研究、原则同意建议、资金来源"进行传译,基本上译出了主讲人欲表达的主题。实践中笔者体会到,在进行多层次的传译时,若译员借助笔记可避免漏译。

四、其他

1.口译时可借助笔记将某些缩略词写出,即使不译,在特定的氛围下,交流双方也能理解其意,如 LOLP(失负荷概率)、GIS(全封闭式组合电器)。

2.由于方言等原因而未听清时,可通过笔记记下来再询问主讲,有助于正确理解原意,确保准确传译。如种植"香蕉树"还是"橡胶树"、是 good 还是 wood 等,写出来可以证实是否听对,否则,凭发音翻译有时会造成译意与说话者原意相差甚远。若遇到类似情况,译员最好还是动动手中之笔。

3.口译涉及面较广,常涉及日常生活、科技经贸等内容,译员不可能对所有专业都清楚,有时碰到生词可请主讲人解释其意,进行解释性翻译,再记录于笔记中便于今后复习。对于实在难以弄懂的词,而且又是关键词,可请主讲人拼写一遍,核查意思后再告知对方。通过笔记记录生词,口译员在实践中可不断积累、逐步提高。

俗话说"好记性不如烂笔头",口译中恰当运用笔记有助于正确理解原意,避免漏译和误译。

第六节　课后练习

(音频:6.6)

Ⅰ. Technical Terms Interpretation

1. 采购计划

2. 延期付款

3. 接受订货

4. 售后服务

5. 备选材料

6. 备选方案投标

7. 港口税费

8. 汇款

9. 汇票

10. 违约罚金

11. EAU (Estimated annual usage)

12. defects liability period

13. advance payment

14. tender security

15. debt due

16. performance security

17. application for export permit

18. countersign

19. letter of indemnity

20. demurrage

Ⅱ. Sentences Interpretation

1. 螺栓(螺钉、螺帽、双头螺栓、弹簧垫圈、销、滚珠轴承、滚柱轴承)是最常用的机械零件。

2. 钢材大致可分为四类,即碳素钢、合金钢、高强度低合金钢和不锈钢。

3. 典型的结构型钢包括工字钢、槽钢、角钢和丁字钢。

4. 采购的第一步是询价,包括在承包人所属国、工程所在国及第三国进行询价。

5. 建筑材料应根据材料的不同性质存放于符合要求的专门材料库房,应避免潮湿、雨淋,防爆、防腐蚀。

Ⅲ. Dialogue Interpretation

【场景:中方机械公司老板(以下简称 A)想要找个厂家包装生产零件,布朗先生(以下简称 B)带领他们参观工厂的相关生产过程。】

A:嗨,布朗先生,我们已参观了工厂,并对你们工厂的生产条件很满意。

B:Yes, they are our main export bases of tools with the advantage of having good production experience and long historical record. All their products enjoy high prestige in the world market.

A:太好了。至于钳子的包装,我了解甚少,我很想知道包装的具体情况。

B:OK, I'll show you how the packing is like. We have a showroom on the ground floor. Let's go downstairs and have a look.

A:可以。

(他们在样品室边走边看包装样品。布朗先生正给客户解释包装情况。)

B:These are the various kinds of packing for pliers. Normally, we have three types of packing: skin packing, hanging packing, and blister packing.

A:这些包装很好看。

B:The skin packing is the most advanced packing for this product in the world market. It catches the eyes and can help push sales.

A:很好,那么出口包装如何?

B：Well, they are packed in boxes of two dozens each, 100 boxes to a wooden case.

Ⅳ. **Passage Interpretation**

Goods procurement refers to the selection of qualified suppliers by contractors through bidding, inquiry, and other forms to purchase materials, equipment, and other goods required in engineering projects. When implementing international engineering goods procurement, it is also necessary to understand relevant information such as international market prices and supply and demand relationships, sources of goods, foreign exchange markets, insurance and transportation, and international trade conditions. Goods procurement management includes the various processes of purchasing or obtaining required materials, equipment, services, or other results from outside the project. These processes include the management and control processes required for the preparation and execution of goods procurement agreements, such as procurement contracts, purchase orders, and memorandum of agreement (MOA), or service level agreements (SLA).

第七章　工程项目管理之进度管理

　　工程项目进度管理是指根据工程项目各阶段的工作内容、工作程序、持续时间和衔接关系编制进度计划，并将该计划付诸实施。在实施的过程中经常检查实际进度是否按计划要求进行，出现了偏差分析原因，采取补救措施或调整、修改原计划，直至工程竣工、交付使用。工程项目进度管理，是施工管理中不可或缺的重要一环。它直接关系到工程能否在合同规定的时间内顺利完工，涉及整个工程的成本和工程各方的利益。施工进度管理是为了保证顺利实施施工进度计划，保证从施工准备到竣工验收的全过程管理和控制。

第一节　背景知识

一、工程项目进度管理标准的设定

　　工程项目进度管理既然以项目的建设工期为管理对象，那么工程项目进度管理的成效就必然由项目建设工期控制的有效程度来体现。由于没有标准也就无所谓控制，因此工程项目进度管理首先要求设定相应的控制标准，这就是目标工期。施工承包企业在目标工期的确定过程中通常有以下几种选择：

　　(1)以预期利润标准确定目标工期；

　　(2)以费用工期标准确定目标工期；

　　(3)以资源工期标准确定目标工期。

二、工程项目进度管理中存在的问题

(一)制约因素多，管理不到位

　　在工程项目实施过程中，影响进度的因素有很多，包括自身的管理水平、施工现场环境、劳动力需求状况、设计变更的影响、资金问题、物资供应问题、风险问题以及其他建设相关方的影响等。工程承包商对这些问题并没有什么积极有效的措施，往往是一个因素产生"共振效应"，从而带动其他因素的影响；管理

组织上不能够保证进度目标的实现,人浮于事、重关系、轻能力现象严重,导致执行能力很差;项目成员只关心自己是否得利,而不管项目目标是否顺利实现;施工项目进行时缺乏有效的监督激励、考核机制以及目标分解不够明确,在进度滞后的情况下找不到直接的负责人,各部门人员之间相互扯皮,最终不了了之;由于没有明确的责任,又缺乏合作精神,项目成员的积极性调动不起来,对进度目标也就不关心。

(二)没有把握好进度、成本、质量之间的关系

工程进度与成本、质量之间是相互联系的。从理论上说,大家都知道成本与进度之间的关系是加快进度就要增加成本。因此,采取赶工措施就要花费一定的费用。进度与质量的关系是加快进度可能会使质量得不到保证。由于人、机械的高强度作业改变了施工条件,质量就可能受到影响。可是,在实际的施工过程中,承包商们并没有花费心思去思考怎样使这三者之间的关系达到均衡。

(三)加强工程项目进度管理的对策

1. 建立符合项目需要的进展状态模板

在启动阶段赢取订单的过程中,承包商通常会与客户签订服务品质保障协议,对项目所应达到的质量做出明确的规定。项目进展状态提交作为其中的一部分,往往也会有原则性的要求。

由于项目进展状态提交的主要目的是向客户概述整个项目的进展情况,因此,项目经理应尽可能与客户采用同一沟通协议,以确保信息能够准确快速地传达。清晰整洁的进展状态模板正是传达沟通协议的具体体现。在明确客户所关注的方面之后,项目经理可以将这些方面具体化到模板上,以便每一次提交都不会遗漏任何信息。

通常,进展状态按周提交,项目经理应在每周固定时间提交或约定时间内将更新后的状态报告提交给客户。报告应如实反映项目当下的最新进展并提出可能存在的问题和需要调整的方面,从而进一步与客户协商解决。

2. 正确反映项目进展状态

项目经理在实施阶段所做的执行与跟踪等工作,正是项目进展状态报告所应涵盖的主要内容。无论是权责体系维护、资源使用管理,还是进度跟踪管理、预算跟踪管理以及范围变更管理,都是项目经理终日忙碌的事情,往往也是客户所关心的问题。毕竟,项目的最终成败同这些方面有直接关系。

项目经理应将实际的项目进展尽可能正确地反映到状态报告中并定期提交给客户。有些项目经理倾向于报喜不报忧,因为担心项目实施过程中出现的问题暴露会降低客户的信任度。然而,这种做法被实践证明是不可取的。须知项目的成功有赖于双方的共同努力,项目实施过程中出现问题也是正常的。而出现了问题却不在第一时间知会所有相关利益方并协力解决,往往会令事态往严重的方向发展,最终导致局势一发不可收拾。所以,正确反映项目实际进展并与客户积极协作,正是提交进展状态报告的本来目的。

第二节　相关词语与表达

序号	中文	英文
1	工期	duration
2	里程碑	milestone
3	网络分析	network analysis
4	箭线网络图	activity-on-arrow
5	节点式网络图	activity-on-node
6	已执行工作实际成本	actual cost of work performed
7	实际开始日期	actual start date
8	实际完成日期	actual finish date
9	行政收尾	administrative closure
10	横道图进度表	bar graph schedule
11	箭线图示法	arrow diagramming method
12	开工	commencement of work
13	逆推计算法	backward pass
14	基准计划	baseline
15	完工预算	budget at completion
16	计划执行预算成本	budgeted cost of work scheduled
17	已执行预算成本	budgeted cost of work performed

续表

序号	中文	英文
18	施工进度	construction progress
19	施工计划	construction plan
20	施工阶段	construction period
21	并行工程	concurrent engineering
22	意外费用	contingencies
23	意外准备金	contingency allowance
24	意外规划	contingency planning
25	合同管理	contract administration
26	合同收尾	contract close-out
27	控制图	control chart
28	纠正措施	corrective action
29	费用预算	cost budgeting
30	费用控制	cost control
31	赶工	hurry
32	关键工序	critical activity
33	关键路线	critical path
34	关键路线法	critical path method
35	当前开始日期	current start date
36	当前完成日期	current finish date
37	数据日期	data date
38	交付物	deliverables
39	挣值法	earned value
40	挣值分析	earned value analysis
41	单节点事件图	event-on-node
42	例外报告	exception report
43	完成日期	finish date
44	职能经理	functional manager
45	职能组织	functional organization
46	甘特图	Gantt chart

续表

序号	中文	英文
47	图解评审技术	graphical evaluation and review technique
48	信息分发	information distribution
49	邀标	invitation for bid
50	关键事件进度计划	key event schedule
51	逻辑图	logic diagram
52	现代项目管理	modern project management
53	蒙托卡罗分析	Monte Carlo Analysis
54	次关键工作	near-critical activity
55	组织分解结构	organizational breakdown structure
56	组织规划	organizational planning
57	整体变更控制	overall change control
58	完成百分比	percent complete
59	执行报告	performance report
60	执行机构	performing organization
61	计划评审技术图	PERT Chart
62	计划开始日期	planned start date
63	计划完成日期	planned finish date
64	优先图示法	precedence diagramming method
65	优先关系	precedence relationship
66	采购规划	procurement planning
67	计划评审技术	program evaluation and review technique
68	项目沟通管理	project communication management
69	项目费用管理	project cost management
70	项目人力资源管理	project human resource management
71	项目综合管理	project integration management
72	项目生命周期	project life cycle
73	项目管理软件	project management software
74	项目管理团队	project management team
75	项目经理	project manager
76	项目网络图	project network diagram

续表

序号	中文	英文
77	项目计划	project plan
78	项目采购管理	project procurement management
79	项目质量管理	project quality management
80	项目风险管理	project risk management
81	项目范围管理	project scope management
82	项目时间管理	project time management
83	质量保证	quality assurance
84	质量控制	quality control
85	请求建议书	request for proposal
86	请求报价单	request for quotation
87	资源约束进度计划	resource-limited schedule
88	资源规划	resource planning
89	责任分配矩阵	responsibility assignment matrix
90	责任图	responsibility chart
91	风险识别	risk identification
92	风险应对控制	risk response control
93	风险应对开发	risk response development
94	进度计划	schedule plan
95	进度计划分析	schedule analysis
96	进度计划压缩	schedule compression
97	进度计划控制	schedule control
98	进度执行指数	schedule performance index
99	进度偏差	schedule variance
100	范围基准计划	scope baseline
101	范围变更	scope change
102	范围变更控制	scope change control
103	范围定义	scope definition
104	范围规划	scope planning
105	范围验证	scope verification
106	询价	solicitation

续表

序号	中文	英文
107	工作人员招募	staff recruitment
108	项目相关者	project stakeholder
109	利润	profit
110	预期利润	expected profit
111	服务酬金	service gains
112	保证	guarantee
113	性能保证	performance guarantee
114	违约罚款保证	penaltiable guarantees
115	机械担保	mechanical guarantees
116	履约保函	performance bond
117	预付款保函	advance payment bond
118	保密协议	secrecy agreement
119	保险	insurance
120	保险单	insurance policy
121	保险证书	insurance certificate
122	保险费	insurance premium
123	信贷额度	line of credit
124	项目融资	project finance
125	剪切裂缝	shear crack
126	贷款	loan
127	索赔	claim
128	培训	training
129	税	tax
130	所得税	income taxes
131	程序	procedure
132	项目协调程序	project coordination procedure
133	项目实施程序	project execution procedure
134	项目设计程序	project engineering procedure
135	项目采购程序	project procurement procedure
136	项目检验程序	project inspection procedure

续表

序号	中文	英文
137	项目控制程序	project control procedure
138	项目试车程序	project commissioning procedure
139	意向书	letter of intent
140	净现值	net present value
141	净现值率	net present value ratio
142	提前完工奖	bonus for early completion of works
143	符合进度表	compliance with schedule
144	同期延误	concurrent delay
145	进度的中断	disruption of progress
146	两班制	double shift
147	提前完工	earlier completion
148	生效日期	effective date
149	可原谅可补偿的延误	excusable compensable delay
150	可原谅但不可补偿的延误	non-compensable delay expedite
151	浮时	float time
152	闲置时间	idle time
153	中间完工日期	interim completion date
154	临时延期的决定	interim determination of extension
155	关键日期	key date
156	劳务计划	labor planning
157	有联系的横道图	linked bar chart program
158	计划编制方法	method of program
159	每年工作日	number of working days per year
160	计划、实施、检查和处理	plan, do, check and action
161	策划日期	plot date
162	计划评估法	program evaluation and review technique
163	进度会	progress meeting
164	暂停工延长	prolonged suspension
165	工作时间的限制	restriction on working hour
166	调整进度表	revise a schedule

续表

序号	中文	英文
167	土木方程进度表	schedule of earthworks
168	竣工报告	statement of completion
169	基本完工	substantial completion
170	工程停工	suspension of work
171	期限	term
172	竣工检验	test on completion
173	竣工时间	time for completion
174	周进度会	weekly progress meeting
175	每周工时分配	weekly time distribution
176	工作命令	working order
177	不可抗拒力	force majeure
178	授标函	letter of acceptance
179	科技英语	English for science and technology
180	专门用途英语	English for specific purposes
181	接手	take possession
182	无效	null and void
183	依据	in terms of
184	基准日	base date
185	解决争议	settlement of disputes
186	达成	enter into
187	法人	legal entity
188	现场监工	inspector of works
189	交流渠道	line of communication
190	永久工程	permanent work
191	主要成本	prime cost
192	工程进度款	progress payment
193	驻场工程师	resident engineer
194	付款条件	term of payment
195	暂定费	provisional sum
196	可行性研究	feasibility study

第三节 对话口译

Dialogue 1:审议和商谈大坝工程的进度（音频:7.3.1）

Word Tips

geological survey and design phase	地质勘测与设计阶段
civil construction phase	土建施工阶段
key milestones	关键节点
construction schedule	施工计划
construction drawings	施工图纸
engineering specifications	工程规范
pouring	浇筑
curing	固化
earthwork engineering	土石方工程
reinforced concrete	钢筋混凝土
quality control	质量控制
safety management	安全管理
installation of large-scale equipment	大型设备安装

【场景:The employer always attaches importance to the project schedule, because if the project is completed earlier, the employer can benefit from the project in a timely manner. Today's meeting is attended by the employer Mr. Reno and the contractor Mr. Liu and Mr. Wang. The aim of today's meeting is to review and discuss the progress of the dam project.

业主始终重视项目进度,因为尽早竣工的话,业主能及时从项目中获得效益。今天参加会议的有业主方 Reno 先生(以下简称 R)和承包方刘先生(以下简称 L)、王先生(以下简称 W),召开会议的目的是审议和商谈大坝工程的进度。】

R:各位早上好。欢迎大家参加今天的会议。我们将审议和商谈大坝工程的进度。刘先生,请你先来汇报一下工程的最新进展。

L:Good morning, Mr. Reno. I'm pleased to report the latest progress of the dam construction project. Currently, we have completed the geological survey and design phase and smoothly transitioned into the civil construction phase. Earlier this month, we finished the foundation treatment work and have started the construction of reinforced concrete structures.

R:非常好,刘先生。请问土建施工是否能按计划进行?还有哪些关键节点需要特别关注?

L:Yes, the civil construction is progressing as planned. We have established a detailed construction schedule and are executing the operations in accordance with the construction drawings and engineering specifications. As for key milestones, we need to particularly focus on the pouring and curing of reinforced concrete, installation of large-scale equipment, and the progress of earthwork engineering.

R:好的,这些都是关键的施工阶段。请问浇筑和固化的进度如何?

L:Pouring and curing is a complex task that requires careful consideration of concrete quality and curing time. We will make every effort to control the construction schedule and ensure the quality and timing of each pouring. Currently, we have completed the first pouring and will proceed with the subsequent pours.

W:除了进度,我们还要关注施工过程中的质量控制和安全管理。我们已经制定了详细的质量控制计划和安全措施,以确保工程的质量和施工场地的安全。

R:These are all very important aspects. I expect everyone to maintain a high sense of responsibility and professionalism, work closely together during the construction process, and ensure the smooth progress of the project. Mr. Liu, is there anything else that needs to be discussed?

L:目前没有其他特别事项,我们将继续按照计划执行工作,并向您及时汇报进展。如果有任何问题或需要调整的地方,我们会及时与您沟通。

R:Thank you, Mr. Liu and Mr. Wang. I am satisfied with the progress of the dam construction project. We will continue to closely monitor the progress and quality control. If there are any matters that need to be discussed, please contact us at any time.

L:Reno 先生,非常感谢您的支持。我们会保持紧密合作,并确保大坝工程按时、优质地完成。

R：OK, this meeting is over. Finally, I wish you all a smooth work and happy cooperation.

Dialogue 2：提交月度报表（音频：7.3.2）

Word Tips

Monthly Statement	月进度报表
materials on site	进厂材料
certify	（工程）签证
variation order	（工程）变更通知
air compressor	空气压缩机
deduction	扣除
retention money	滞留金

【场景：The project manager Mr. Bai shall submit the Monthly Statement to the owner Mr. Ai.

项目经理白先生（以下简称 B）向业主艾先生（以下简称 A）提交月度报表。】

B：今天 28 号，是提交月进度报表的日子。

A：Yes it is. Please hand it in if you have it with you.

B：给您，这上面显示了到本月 25 日的已完成工程量。这些已根据现场进度和图纸核实过了。

A：(studying) Fine. Where do these unit prices come from?

B：它们来自我们投标文件中的工程量清单。

A：Well, your statement is OK in general, but Item 1, the concrete foundation for air compressors should be canceled from the list because they didn't pass our inspection two days ago. There are several cracks on the concrete surface. You should remove and reconstruct them immediately on your own cost.

B：好的。

A：What is this item for?

B：这是本月进场的材料和设备量。

A: But according to the Contract, the sum in the Monthly Statement for materials on site should be only 70% of the value of the materials and equipment delivered to site.

B:是的,我们的确根据合同从月进度报表中去掉了材料和货物价值的30%。请签证本月变更通知单上的工程量。

A: We can certify the quantity now, but the prices of some variation items seem to be quite high. What is your pricing based on?

B:这些报价基于我们材料供应商的报价,因为这些材料不同于原合同中的材料而且它们的单价无法根据工程量单确定。

A: Your pricing in this part needs to be negotiated further. Let's leave it aside for our next discussion.

B:好的。您还有其他意见吗?

A: I'd like to remind you that the amount of your Monthly Statement is subject to a 10% deduction of retention money.

B:谢谢您的提醒。我明天早上就给您修改后的月进度报表。

第四节　篇章口译

Passage 1:工期滞后原因分析（音频:7.4.1）

Word Tips

construction contract requirements	施工合同要求
cross operations	交叉作业
partition wall	隔断墙
single sided gypsum board	单面石膏板
suspended ceiling keel frame	吊顶龙骨架
total heat exchange fan	全热换风机
air tightness test of air conditioning refrigerant pipes	空调冷媒管气密试验
fire sprinkler pipes	消防喷淋管

【场景:项目经理刘鹏在工作进度会议报告上分析工期滞后的原因。

Project manager Liu Peng analyzes the reasons for the delayed construction period in the work progress meeting report. 】

Good morning, everyone. The entire project has been under construction for 105 working days since February 10th. According to the construction contract requirements, the completion of the entire project should be on May 10th, and it has been over 15 days so far. Based on the analysis of the preliminary construction situation, the reasons for the delay in the construction period are roughly as follows:

First of all, the general contractor of the construction unit is not clear about the disclosure of the operators, ineffective control of the professional construction team, there is no corresponding restriction measures, and the cross operation arrangement is not in place, resulting in repeated construction in some parts. After the single gypsum board sealing of the partition wall of building 3, the sound insulation materials were not installed in time. In the regular meeting, the construction unit was repeatedly required to complete the installation as soon as possible, but the implementation effect was not ideal. At present, in the ceiling of building 2, fire fighting and weak current pipelines are laid, and the ceiling keel frame has been installed in place. But the full-heat exchange fan, used for air supply and exhaust, has just arrived. These conditions have caused a great impact on the progress of the project.

Secondly, the air ducts on each floor have been basically installed and laid. The air tightness tests for the air conditioning internal units and refrigerant pipes on each floor were completed on May 5th. Currently, the air conditioning external units in buildings 2 and 3 have not yet arrived. The external units need to be lifted and placed on site, and the system connection, vacuum testing, and debugging of the external units also require some time. This is also an important factor affecting the progress.

In the end, the issue of fire safety construction approval is also one of the reasons that has affected the progress of buildings 2 and 3. The proof of fire protection equipment was only received on March 24th for this renovation project. Previously, in order to not affect the project progress, the owner and property management agreed to remove the fire sprinkler pipe at the sixth engineering meeting on March 18th.

Passage 2：实现进度目标的举措（音频：7.4.2）

Word Tips

project schedule control	项目进度控制
schedule goals	工期目标
design drawings	设置图纸
flexible and dynamic control	灵活动态管控
rolling plan	滚动计划
baseline plan	基准计划
idle work and rework	窝工、返工

【场景：项目经理王先生在工作进度会议报告上阐明今后实现进度计划目标的举措。

Mr. Wang, the project manager, explained in the work progress meeting report the measures to achieve the schedule goals in the future. 】

大家早上好，我是项目经理王鹏。接下来，我将就今后实现进度计划目标的举措进行发言。

首先，我方在承接贵公司国际重大工程项目时，将采用国际上最先进的 BIM 标准。在项目的全生命周期，我方将运用 30 多款 BIM 软件来优化整个项目管理工作，为项目进度控制提供了保障。

其次，我方将根据项目实施方案的工期目标、资源配置能力、设计图纸提供情况以及外部条件等因素，编制项目进度计划，再将具体项目进度计划进一步分解为周计划和日计划，从而实现进度计划目标以周保月，以月保年。

然后，在施工过程中，我方将实行灵活动态管控。在调整施工项目过程中，业主通常要求承包商提交 2 周滚动计划、施工进展周报等文件以证明并保证项目进度满足合同要求。对此，我方保证，我方将及时对比项目的基准计划和过程中的跟踪计划，一旦发现偏差，及时分析问题原因，根据现场实际情况，采用多种手段调整分项计划，使实际工期与基准工期计划匹配。

最后，我方将严控安全质量。我方将始终秉持"安全第一，质量至上"的原则，在项目进度管理中通过优化制度流程、增加项目考评等手段加强对项目安全质量的监督、检查与指导，避免窝工、返工现象的发生。

谢谢大家！

第五节 口译技能——数字口译技巧

随着全球经济一体化的快速发展,中外交流日益频繁,口译需求不断增加,对口译人员的翻译要求也在提高。在商务谈判等各种翻译场合,数字口译不可避免地经常被使用。由于中文与英文数字计数方法不同,数字口译有非常多的特殊性,也会给翻译人员造成阻碍。那怎么来提升数字口译的技巧呢?

一、英语和汉语在数字表达上的差别

(一)数字位数不同

在中文里,每一个数字都有自己的表达方式,从小到大,个、十、百、千、万、十万、百万、千万……这种数字表达简单易记。英语的数字表达却与汉语大不相同,百位数以上的数字,一般以三位数为一个计量单位,顺序为百(hundred)、千(thousand)、百万(million)、十亿(billion),所有其他数字需要按照这四个基本单位进行计算。在中文里有"万"这个计数单位,英文中没有"万"这一计数单位,英语中的"一万"为 ten thousands。在数字翻译过程中要注意中文与英语数字的表达方式,中文与英语数字表达差异最大的是在"十亿"这个数字上。

如中文的 60 亿,翻译成美式英语则为 six billion,翻译成英式英语则为 six thousand million;如 800 亿,翻译成美式英语则为 eighty billion,翻译成英式英语则为 eighty thousand million。从以上举例可以看出,英式英语对 billion 的表达方式与美式英语不同。在英语口译中要区分美式英语与英式英语的差别,减少翻译过程中的错误。

(二)数字单位不同

汉语中常见的单位有亩、尺、斤,但是英文中没有这些单位,而是用 gram、kilometer、inch、mile 等来表达。单位不同也会给翻译人员带来阻碍,翻译人员要正确掌握这些单位,进行准确的换算后再进行翻译。

(三)倍数表达方式不同

中文和英文的倍数表达方式也有差异,在中文中倍数可以用两种形式进行表达,例如 A 是 B 的几倍,B 比 A 大几倍。这两种表达方式表达的意思是不同的,而在英文中这两种表达方式是一样的。如:中文的"B 是 A 的 4 倍"表达的意思是 1 个 B 等于 4 个 A,而"B 比 A 大 4 倍"表达的意思是 1 个 B 等于 5 个 A;

英语则不同,英语中 B 是 A 的 4 倍可以表达为"B is four times as large as A",也可以表达为"B is four times larger than A",两种表达意思是一样的。

二、英语口译中数字口译的技巧与方法

(一)数字的正确记录法

数字本身是比较难记的,需要正确记录,以保证口译工作的顺利进行。口译人员在翻译数字时要保证准确、快速,就要先有技巧地准确记录。

首先,简单的缩略记录。缩略记录主要用于特别复杂的英文数字的表达,用阿拉伯数字与单位进行简写来表达。如在记录"eighty-eight million three hundred and fifty-four thousand three hundred and thirteen"时,口译者记作"88m354t313"。这是一种简单的转换记录法,不适合换算数字的口译记录。

其次,特殊符号记录法。英语的计数符号主要有 4 种,从千位开始进行转换,可以用符号进行隔开,隔开符号常用逗号。在记录时可以用一个逗号表示千,两个逗号表示百万,三个逗号表示十亿,例如 six billion 可以记录成"6,,,"。

再次,填充记录法。填充记录法也是针对英语的 4 个基本计数单位设计的,先将计数单位写在纸上,口译时直接填充数字。填充记录法简单,不易出错。

(二)单位的翻译

在进行英语口译时一些数字不是单独出现的,往往带有单位。在口译时需要先准确翻译单位,再进行数字翻译,以保证翻译的精准性。因为在口译过程中单位出现错误,表达的意思就是错误的。口译者一定要重视单位的翻译,避免因单位翻译错误导致整个翻译错误,要记住面积、长度、重量和体积的正确翻译方法,一些常见单位可简写,以保证翻译正确。

(三)倍数的翻译

中文与英文在倍数表达上差异特别大。中文的"B 是 A 的 4 倍",英文翻译是"B is four times as large as A"或"B is four times larger than A"。由于表达方式不同,在最初口译时尽量翻译成"B is four times as large as A",避免错误发生,提高数字口译的精准性。

总之,在英语口译时要加强对数字口译的重视及正确认识,在记录方法、单位翻译、倍数翻译方面都保证正确。

第六节 课后练习

（音频:7.6）

Ⅰ. Technical Terms Interpretation

1. 不可抗拒力

2. 两班制

3. 土木工程进度表

4. 开工

5. 意外费用

6. 赶工

7. 关键路线法

8. 信息分发

9. 质量保证

10. 责任分配矩阵

11. projector stakeholder

12. expected profit

13. schedule variance

14. near-critical activity

15. percent completion

16. cost budgeting

17. baseline

18. term

19. weekly progress meeting

20. statement of completion

Ⅱ. Sentences Interpretation

1. 我们应该按照工程项目的总进度表(建设进度表)工作。

2. 卖方将于五月份向买方提供初步(最终)技术文件。

3. 在我们的大多数工程项目中都采用"统筹法"(即关键路线法,简称

CPM)安排计划。

　　4.卖方操作组(专业工作组)将在现场,一直工作到生产符合保证条件。

　　5.每个月我们都要制定建设进度计划。

　　6.这座厂的交工验收期将在 2025 年 4 月 6 日。

　　7.这座合同工厂将于 2028 年 1 月 1 日试运转。

　　8.我们主要的计划工作项目包括成本估算和施工进度。

Ⅲ. Dialogue Interpretation

　　【场景:业主安先生(以下简称 A)在工程月进度会上和项目经理李先生(以下简称 L)对话。】

　　A:非常遗憾,我必须说对你们的施工进度不太满意。根据厂房改建工程的总进度计划,你们本应该现在就已经完成屋面安装工程了。

　　L:It was delayed by the bad weather. You see, the site diary shows there were 5 days with heavy rain and 2 days with strong wind. We had to stop roof installation in those days.

　　A:主要工艺设备必须在下个月 10 号开始安装,这是关键路径上的关键工作。如果你们无法满足这个关键日期的要求,将会严重影响竣工和移交。所以,你们必须完成。合同规定的关键日期是没有商量余地的。

　　L:I see.

　　A:而且,你们的暖通设备安装进度似乎有些落后。

　　L:It was held up because you delayed the approval of equipment procurement.

　　A:我不同意。你们希望用不同的设备来代替技术说明书中规定的设备。然而,直到上个月 15 日我们才收到你们的书面报告。在仔细地研究你们建议的设备后,我们于上个月 26 日给你们发了我们的确认书。我们的回复仍在合同规定的 2 周期限内,这是我们的文件传送记录。所以,不用再谈这个问题了。现在,重要的是在尽可能短的时间里完成该项工作。

　　L:We can finish it within 50 days.

　　A:不行,太迟了。你们能采用两班制在 45 天内完成吗?

　　L:We will try our best. We'll mobilize more resources and re-schedule the works. I promise you that the modified schedule will be submitted to you within 3 days. Another thing I'd like to mention is that we want to apply for a time extension and cost additional for the earthwork of the Additional Production Building. Your revised layout was supposed to be issued a week ago. But we haven't received it yet. Even if we get it today, we could hardly catch up with the normal schedule without additional manpower.

A:关于平面图还有一些问题未能确定,我们会加快速度的。但是推迟关键时间是不可能的,因为每个关键日期都与下一个日期相关联。我们同意考虑增加费用,但你们首先应该提交书面申请。

L:OK, we will prepare it.

IV. **Passage Interpretation**

【场景:该段为施工人员汇报隧道挖掘的进度情况。】

Let's start with the diversion structure. For the cofferdam, we have placed 2,600 cubic meters of concrete. It will be finished within the next few days and on schedule. At the left abutment and the dam foundation, we have completed 5,000 cubic meters of rock excavation. That's what had been achieved at the diversion structure during the preceding month. Let's now have a look at the tunnel excavation. At Adit A, we have advanced 120 meters at the upstream working face and 80 meters at the downstream. We've also done 590 pieces of steel rock bolt which amount to 4.5 tons. At Adit B, the tunneling advance is 40 meters. Rock bolt, 750 pieces. The portal of Adit B adds up to 15 tons. That's for the tunneling. Now, the power house. The foundation excavation is being done there. The quantity completed for the past month is 5,000 cubic meters. And last, the installation of the stone crushing plant has been completed and it is now test running.

第八章　工程项目管理之质量管理

质量是指"一组固有特性满足要求的程度"。质量概念的关键是"满足要求"，这些"要求"必须转化为指标，作为评价、检验和考核的依据。另外，质量要求是随时间、地点、环境的变化而变化的，是动态的、发展的和相对的。

因此，应定期评定质量要求、修订规范标准，不断开发新产品、改进老产品，以满足新的质量要求。在国际工程承包中，不同的国家、不同的地区会因自然环境条件、技术发展水平等的不同而对工程质量提出不同的要求，工程承包商应该适应这种区域差异产生的不同要求，根据不同的要求提供不同性能的工程。

第一节　背景知识

施工安全是各行业工程建设中必需的保障。施工安全涵盖了在作业过程中所有的安全问题并且涉及管理、财务及后勤保障等相关内容。

广义的工程质量管理，泛指建设全过程的质量管理，贯穿于整个工程建设，涵盖决策、勘察、设计、施工的全过程。

一般意义的质量管理，指的是工程施工阶段的管理。专家认为质量管理应从系统理论出发，把工程质量形成的过程作为整体。世界上许多国家要求工程项目以正确的设计文件为依据，结合专业技术、经营管理和数理统计，建立起一整套施工质量保证体系，才能投入生产和交付使用。因此，要采用最经济的手段，根据质量标准，引入科学的方法，对影响工程质量的各种因素进行综合治理，建成符合标准、用户满意的工程项目。

工程质量管理，要求把质量问题消灭在它的形成过程中。工程质量管理要以预防为主，保证手续完整，并在全过程、多环节致力于质量的提高。这就要求把工程质量管理的重点，从"以事后检查把关为主"变为"以预防、改正为主"。

组织施工要制定科学的施工组织设计,从管"结果"变为管"因素",把影响质量的诸因素查找出来,依靠科学理论、程序、方法,发动全员、全过程、多部门参加,保证参加施工的人员均不遭受重大伤亡事故,使工程建设全过程都处于可控范围内。

一、决策阶段的质量管理

此阶段质量管理的主要内容是在广泛搜集资料、调查研究的基础上研究、分析、比较,论证项目的可行性,确定最佳方案。

二、施工前的质量管理

施工前的质量管理的主要内容有:

1. 对施工队伍的资质进行重新审查,包括各个分包商的资质审查。如果发现施工单位与投标时的情况不符,必须采取有效措施予以纠正。

2. 对所有的合同和技术文件、报告进行详细审阅,如图纸是否完备,有无错漏空缺,各个设计文件之间有无矛盾之处,技术标准是否齐全。应该重点审查的技术文件除合同外,还包括审核有关单位的技术资质证明文件;审核开工报告,并经现场核实;审核施工方案、施工组织设计和技术措施;审核有关材料、半成品的质量检验报告;审核反映工序质量的统计资料;审核设计变更、图纸修改和技术核定书;审核有关质量问题的处理报告;审核有关应用新工艺、新材料、新技术、新结构的技术鉴定书;审核有关工序交接检查报告、分项和分部工程质量检查报告;审核并签署现场有关技术签证、文件等;审核是否配备检测实验手段、设备和仪器;审查合同中关于检验的方法、标准、次数和取样的规定。

3. 审阅进度计划和施工方案。对施工中将要采取的新技术、新材料、新工艺进行审核,核查鉴定书和实验报告。

4. 对材料和工程设备的采购进行检查,检查采购是否符合规定的要求并协助完善质量保证体系。

5. 对工地各方面负责人和主要的施工机械进一步审核。

6. 做好设计技术交底,明确工程各个部分的质量要求。

7. 准备好简历、质量管理表格,做好担保和保险工作。

8. 签发动员预付款支付证书。

9. 全面检查开工条件。

三、施工过程中的质量管理

1. 工序质量控制

具体包括施工操作质量和施工技术管理质量。

①确定工程质量控制流程；

②主动控制工序活动条件，主要指影响工序质量的因素；

③及时检查工序质量，提出对后续工作的要求和措施；

④设置工序质量控制点。

2. 设置质量控制点

对技术要求高、施工难度大的某个工序或环节，设置技术和监理的重点，重点控制操作人员、材料、设备、施工工艺等；针对质量通病或容易产生不合格产品的工序，提前制定有效的措施，重点控制；对于新工艺、新材料、新技术也需要特别重视。

3. 质量检查

包括操作者的自检、班组内互检、各个工序之间的交接检查、施工员的检查和质检员的巡视检查、监理和政府质检部门的检查。具体包括：

①装饰材料、半成品、构配件、设备的质量检查，并检查相应的合格证、质量保证书和实验报告；

②分项工程施工前的预检；

③施工操作质量检查，隐蔽工程的质量检查；

④分项分部工程的质检验收；

⑤单位工程的质检验收；

⑥成品保护质量检查。

4. 成品保护

①合理安排施工顺序，避免破坏已有产品；

②采用适当的保护措施；

③加强成品保护的检查工作。

5. 交工技术资料

交工技术资料主要包括以下文件：材料和产品出厂合格证或者检验证明、设备维修证明；施工记录；隐蔽工程验收记录；设计变更、技术核定、技术洽商；水、暖、电、声讯、设备的安装记录；质检报告；竣工图、竣工验收表等。

6.质量事故处理

一般质量事故由总监理工程师组织进行事故分析,并责成有关单位提出解决办法。重大质量事故,须报告业主、监理主管部门和有关单位,由各方共同解决。

7.工程完成后的质量管理

按合同的要求进行竣工检验,检查未完成的工作和缺陷,及时解决质量问题,制作竣工图和竣工资料。维修期内负责相应的维修责任。

第二节 相关词语与表达

序号	中文	英文
1	现场工程师	resident engineer
2	顾问工程师	advisory engineer
3	总工程师	chief engineer
4	咨询工程师	consulting engineer
5	工区工程师	division engineer
6	土木工程师	civil engineer
7	电气工程师	electrical engineer
8	机械工程师	mechanical engineer
9	岩土工程师	geotechnical engineer
10	放松,松懈	slacken
11	坍落度试验	slump test
12	浇筑现场	placement site
13	水灰比	water/cement ratio
14	裂缝与凸面	breaks and steps
15	蜂窝	honeycomb
16	夯实与振捣	compaction and vibration
17	管网布置	multiple runs
18	电缆预埋管	embedded electrical conduit
19	集料	aggregate
20	弱结合面/层	weakened plane
21	补救措施	remedy

续表

序号	中文	英文
22	有缺陷的混凝土	defective concrete
23	风钻	pneumatic drills
24	施工详图	working drawings
25	全面质量管理	TQM
26	拨款	grant
27	周转性贷款	revolving loan
28	恢复,复位	restoration
29	手艺,工艺	workmanship
30	周期的,定期的	periodic
31	过剩水量	excess water
32	返工	rework
33	陈旧技术	antiquated techniques
34	供应链不足	supply-chain deficiencies
35	规格,说明书	specification
36	质量管理	quality management
37	建设项目	construction project
38	建筑行业	construction industry
39	设计加工	design process
40	高额维修费用	costly repairs
41	项目工程师	project engineer
42	竣工验收	completion acceptance
43	结构设计	structural design
44	维修费用	maintenance costs
45	结构强度	structural strength
46	加工工业	process industry
47	公共设施系统	utility system
48	基金资助机构	funding agency
49	系统经理	system manager
50	特殊条款	specific item
51	循环贷款	revolving loan

续表

序号	中文	英文
52	分类财政补贴	block grants
53	合理选材	proper material
54	许可证颁发机构	permit agency
55	第三方	a third party
56	工程质量管理	CQM
57	资格预审	prequalification
58	清偿,结算	liquidate
59	滞留,保留	retention
60	保修期	maintenance
61	环境影响评估	EIA
62	无异议证书	NOC
63	质量保证	QA
64	质量控制	QC
65	碳酸化	carbonation
66	磨损	abrasion
67	策划,实施,检查,处置	plan, do, check, action
68	附属设施	ancillary facilities
69	产品质量先期策划	advanced product quality planning
70	评审小组	review panel
71	统计过程控制	statistical process control
72	施工图预算	construction drawing budget
73	控制计划	control plan
74	质量体系评定	quality system assessment
75	每百万零件不合格数	parts per million
76	质量手册	quality manual
77	质量程序文件	quality procedure
78	过程能力指数	process capability index
79	机器能力指数	machine capability index
80	设备综合效率	overall equipment efficiency
81	质量功能展开	quality function deployment

续表

序号	中文	英文
82	顾客导向过程	customer oriented processes
83	质量管理体系	quality management system
84	质量审核	quality audit
85	成品质量控制	final quality control
86	进货质量控制	incoming quality control
87	过程质量控制	process quality control
88	出货质量控制	outgoing quality control
89	可记录工伤值	total record injury
90	竣工结算	completion settlement
91	潜在失效模式及后果分析	potential failure mode and effects analysis
92	安全预评价审批	safety pre-assessment approval
93	测量系统分析	measurement system analysis
94	作业制成本制度	activity-based costing system
95	保密设施	confidential facilities
96	可承诺量	available to promise
97	产能需求规划	capacity requirements planning
98	客制化生产	configuration to order
99	成熟度验证	design maturing testing
100	项目立项	project authorization
101	企业流程再造	business process reengineering
102	成品质量管理	finish or final quality control
103	首批样品认可	initial sample approval request
104	最小总成本法	least total cost
105	最小单位成本	least unit cost
106	委托代工	original equipment manufacture
107	在线分析处理	online analytical processing
108	PDCA 管理循环	plan-do-check-action management cycle
109	产品数据管理系统	product data management system
110	预估在手量	product on hand
111	退货验收	returned material approval

续表

序号	中文	英文
112	特殊订单需求	special order request
113	供应链管理	supply chain management
114	现场控制	shop floor control
115	策略信息系统	strategic information system
116	全面生产性维护	total productive maintenance
117	统计制程管制	statistic process control
118	在制品	work in process
119	生产线品质控制	line quality control
120	供应商品质保证	source(supplier) quality assurance
121	制程品质保证	process quality assurance
122	严重缺点	critical defect
123	工程改动要求	engineering change order
124	兼容性	compatibility
125	首件检查报告	first piece inspection report
126	核价流程	price verification process
127	方案设计	conceptual design
128	发布人事任命通知	issue personnel appointment notice
129	全面品质管理	total quality management
130	制程不良分析	manufacturing defect analysis
131	纠正措施需求	corrective action request
132	允收	ACC
133	拒收	REJ
134	抽样检验样本大小	sample size
135	制造专案管理	manufacturing project management
136	质量交付成本	quality delivery cost
137	主要绩效指标	key performance indicate
138	制造验证试产	manufacturing verification test
139	产能改善计划	capacity improvement plan
140	物料审核小组	material review board
141	退货单	material reject bill

续表

序号	中文	英文
142	制程检验规格	specification inspection process
143	测试作业流程	test operation process
144	故障分析报告	failure analysis report
145	表面贴装器件	surface mounting device
146	弹性制造系统	flexible manufacture system
147	生产件批准程序	production part approval process
148	自然条件	natural conditions
149	设备总效率	overall equipment effectiveness
150	计算机辅助能力设计	computer-aided design
151	质量功能展开	quality function deployment
152	支持性过程	support process
153	过程流程图	process flow chart
154	企业需求计划	enterprise requirement planning
155	作业基础预算制度	activity-based budgeting system
156	先进规划与排程系统	advanced planning and scheduling system
157	应用程序服务供货商	application service provider
158	认可的供货商清单	approved vendor list
159	平衡记分卡	balanced score card
160	计划生产	build to forecast
161	要径法	critical path method
162	限制驱导式排程法	drum-buffer-rope
163	基本经济订购量	economic order quantity
164	国际标准组织	International Organization for Standardization
165	逐批订购法	lot-for-lot
166	限制理论	theory of constraints
167	气象资料	meteorological data
168	最佳生产技术	optimized production technology
169	制程能力	capability index of process
170	准确度	capability index

第三节　对话口译

Dialogue 1：施工现场安全管理（音频：8.3.1）

Word Tips

Safety Officer	安全主任
Safety Scheme	安全计划
helmet	安全帽
hard shoes	劳保鞋
Work Permit	工作许可证

【场景：Safety Director A is conducting an on-site quality check on the implementation of the welding project, and Welder W is being inspected.

安全总监 A 正在对焊接项目的实施进行现场质量检查，焊工 W 正在接受检查。】

A：I'd like to speak to this welder first. Did you have any training courses for safety education?

W：是的。在课堂上，我学习了所有与焊接相关的安全规定。

A：It's fine that you wear the helmet and protective goggles, but where are your hard shoes? Didn't your manager give one pair to you?

W：是的。对不起，我今天忘记穿了。

A：Don't forget again. Did you get a "Work Permit" for the welding here?

W：当然有，在这里。

A：(by an opening) I can see from here that there are neither guard rails nor sufficient covers for such a big opening in the floor. It does not comply with the safety regulations and should be corrected right now.

W：(对工头说)工头，你要马上安排两个工人把洞口封好。

A：Also, remember to set a warning notice here to alert people of the hazard in this area.

W：我们将立即按照您的指示进行处理。

A：OK, that's all for the safety patrol today. I hope all the non-conformances will be corrected soon and will not be repeated. I'll check them one by one in the next safety patrol. Thank you.

Dialogue 2:工程细节核查（音频:8.3.2）

Word Tips

地脚螺栓	foundation bolt
扭力矩	the moment of twisting
垫铁组	pad-iron groups
图纸规范	drawing standard

A:Have you completed the installation of this pump?

B:是的,已经完成了。

A:How about the circumstances?

B:很好！地脚螺栓垂直,螺栓拧紧而且扭力矩一致,螺母与垫圈、垫圈与设备底座的接触很紧密。

A:How about the pad irons?

B:垫铁组平稳,接触紧密。每组不超过三块。

A:Please explain how many millimeters is the requirement of the coupling butt end space?

B:最小间隙达到20毫米即可。

A:The elevation of this equipment installation is wrong, isn't it?

B:不,标高符合图纸规范的要求。

A:Excuse me, whether this compressor needs disintegration for check-up?

B:是的。

A:May I set the machine going?

B:不,还要检查一下。

Dialogue 3:检查安装任务（音频:8.3.3）

Word Tips

蒸馏塔	distillation tower
电动绞车	electrical winch

滑轮   pulley

A:诸位,早安。

B:Good morning.

A:今天,我们的任务是安装这台蒸馏塔,请准备。

B:Yes. Tell us the weight, length, diameter and center of gravity of this tower, please.

A:这塔的重量是 34 吨,长度 43 米,直径 9 米,重心在离塔底 15 米处。

B:Who is the director of the installation work?

A:王工程师今天负责安装工作,彻底检查一下工具,如电动绞车、钢丝绳、滑轮,弄清楚它们是否都处于良好状态。进一步与图纸对照,检查一下基础螺丝(地脚螺栓)。

B:OK, we have checked over them all.

A:好,请开始工作吧!

W:Pay attention everyone! Listen to my whistle on your post.

A:好了! 该塔已就位了,检查一下它的垂直度。

W:The perpendicular tolerance of the tower is less than one thousandth of its height.

B:Acceptable!

A:好,请拧紧基础螺帽。休息吧。

第四节　篇章口译

Passage 1（音频:8.4.1）

Word Tips

client satisfaction   客户满意度

robust risk mitigation strategy   稳健的风险缓解策略

【场景:安全总监李飞正在向委托方介绍工程质量管理的重要性以及必要性。】

Good day, ladies and gentlemen. Let me introduce myself first. My name is Li

Fei. Today, I am honored to speak to you about the crucial aspects of engineering quality management, highlighting its importance and necessity in the field.

Engineering quality management is an indispensable component of any successful project. It encompasses a set of systematic processes and practices aimed at ensuring that a project meets the specified quality standards and requirements. This discipline plays a pivotal role in the overall success and longevity of engineering projects across various industries.

One of the primary reasons why Engineering quality management is essential lies in its ability to ensure compliance with established standards and regulations. In today's globalized world, projects often involve collaboration among diverse stakeholders, each with their own set of guidelines. Effective quality management guarantees that a project aligns with these standards, fostering international cooperation and minimizing the risk of legal complications.

Quality management acts as a robust risk mitigation strategy. Identifying potential risks and implementing proactive measures to address them are integral parts of this process.

By systematically managing risks, engineering projects can avoid costly mistakes, delays, and reputational damage. This proactive approach enhances the overall resilience of a project in the face of unforeseen challenges.

Investing in quality management upfront can yield significant cost savings in the long run. By identifying and rectifying quality issues during the early stages of a project, organizations can prevent the need for costly rework and modifications later on. This not only saves financial resources but also ensures that the project adheres to its budget and timeline.

Engineering quality management directly contributes to client satisfaction. Meeting or exceeding quality expectations enhances client satisfaction.

Passage 2 (音频:8.4.2)

【场景:在A公司工作的项目质量管理代表王健正在向客户们介绍公司业务和已完成的典型案例。】

早上好！我是王健,在 A 公司工作,我是工程项目质量管理代表。首先,感谢你们给我公司提供了这次机会。接下来,我想向你们介绍一下我公司的业务。

我们的业务涵盖了工程质量管理的各个方面,包括但不限于:通过严格的检查和监控程序,确保项目在每个阶段都符合相关标准和规范;提供全面的质量保证服务,以确保项目的整体质量达到客户的期望并超越预期。采用先进的技术和创新方法,结合行业最佳实践,以确保我们的客户获得最佳的工程质量管理服务。

我们致力于不断改进和创新,以适应不断变化的行业需求和标准。我们公司在前年负责了江西的一座水电站项目的质量管理,项目圆满交付,客户对我们赞不绝口,并且该项目运行两年多以来,从未出现安全事故或发电故障。这不仅展示了我们的专业能力,也证明了我们对质量和客户满意度的承诺。

总的来说,我们公司以专业、可靠和创新的质量管理服务著称,致力于帮助客户确保工程项目达到高质量标准并取得成功。我们将在招标书后附上关于我公司详细信息的一些手册。您可以通过这些手册详细了解我公司总部的详情,以及我公司过去承接的多个项目质量管理案例。

第五节　口译技能——语言重组

翻译不是在两种语言中寻找字词或句法的完全对应,而是追求信息的对等。但是由于时间和心理上的巨大压力,口译员在工作时容易受源语的限制,在译文中常会生搬硬套源语的表达和结构,使译文不符合目的语的习惯,有时甚至不知所云。如有的同学听到"We started this magazine to write about people who make things happen in business.",便会译成:"我们创办这本杂志是写那些使商界事情发生的人。"

什么是"使事情发生的人"？其实就是商界的开拓者、改变商业社会的人。因此译员听到一段讲话后应尽量脱离源语的结构,概括其含义,用符合目的与习惯的方法将信息重新组织。

由于汉英两种语言的巨大差异,有些表达方式直译过来并不能在目的语中产生预期的效果,甚至有时让一些英语国家的人产生误解和不悦。有的字词在目的语中没有意义对等的字词,需要进行必要的解释,有的讲话人因为过于紧张或者准备不足而啰唆、重复,词不达意,甚至说到一半就卡壳。这种情况下译员应对讲话进行必要的整理和补充,以协助讲话人传递信息。

一、英汉两种语言的特征对比

美国的著名语言学家、翻译家尤金·A.奈达在《译意》中指出,汉语和英语在语言学上最重要的区别就是形合与意合的对比。

英语是"显性连接",语言组织主要靠语言本身的语法手段,即句子内部的连接或句子间的连接采用句法手段或词汇手段。英语重"形合",语句各成分的相互结合常用适当的连接词语或各种语言连接手段,以表示其结构关系。

汉语语言组织主要靠语言本身的语义手段,即句子内部的连接或句子间的连接采用语义手段。汉语重"意合",句中各成分之间或句子之间多依靠语义贯通,少用连接语,所以句法结构形式短小精悍。汉语是"隐性连接"。

让我们来看一个中文例子:

"一只青蛙一张嘴,两只眼睛四条腿。"

"A frog has a mouth, two eyes and four legs."

再看一个英文例子:

"The boy had his breakfast and went off to school."

本句中"his"跟"the boy"前后照应,"and"将两个承接的动作衔接起来,是英文形合的手段。如果翻译时过于拘泥于原文的"形合",译文("男孩吃过他的早饭,然后上学去了")会显得累赘,而汉语只要说"男孩吃过早饭上学去了"即可,两个连贯的动作显得很紧凑,说出的话一气呵成。让我们再看一句:

"我在美国队进入会场的时候看到你穿上西装站起来向他们招手。"

这个句子语句连贯,一气呵成,是典型的中文流水句。而译为英文时却要大量使用连接词,才符合英文的语法规则:

"When the U.S. team marched into the stadium, I saw that you put back your coat and your suit and then you stood up and cheered for your team."

总而言之,汉译英时要先分析汉语句子的功能、意义,以及隐含的逻辑关系,再确定句子的结构、形式,适当借助关联词语连接;英译汉时则要先分析英语句子的结构、形式,再确定它的功能、意义,在意义上体现逻辑关系,在形式上尽量简洁明了,省略冗余词语。

二、语言转换的原则

1. 在内容上力求准确,符合源语的真实意思。

2. 在形式上做到通顺,符合目的语的语言习惯。

三、语言重组技巧

1. 转换:包括词性转换、句型转换和语态转换。词性转换如英语中的名词和介词译成汉语时可以变为动词,汉语中的某些动词译成英语时可以变成名词或介词。句型转换指将句中成分的语序进行调整,例如调整中心词与修饰限定词的位置。例:

我代表…… On behalf of. . .(动词变介词短语)

He is a good speaker. 他擅长演讲。(形容词变动词)

2. 释义:摆脱源语句式,以符合目的语习惯的表达将原意表述出来。

3. 增补与省略:增补是指将发言者未能表达的意思说出来,或面对非专业听众将某些术语或习语进行解释。而省略是指将重复冗余的信息或啰唆的话省略。增补和省略不意味着自由发挥,而是为了更好地将原文的实际意思用符合目的语习惯的表达口译出来,要做到增减词不增减意。

4. 拆分与合并:长句的拆分和短句的合并。

5. 顺句驱动:指按照与源语相同的语序或表达方式把句子译成目的语。它是同声传译中最重要的语言转换技巧,在交替传译中也经常使用。

以下是联合国第八任秘书长潘基文在南京大学演讲时的开场白:

I am delighted to be here on this beautiful campus at one of the world's oldest centers of learning.

很高兴能来到这里,来到这个美丽的校园,这是世界最古老的学府之一。

第六节　课后练习

（音频:8.6）

Ⅰ. Technical Terms Interpretation

1. shop floor control

2. yield point

3. work in process

4. tensile strength

5. field inspection work

6. elongation percentage

7. process quality assurance

8. Brinell hardness

9. flexible manufacture system

10. job-built

11. 制造验证试产

12. 压槽

13. 全面生产性维护

14. 集水板

15. 缺陷分析系统

16. 安全主任

17. 制造专案管理

18. 镶边区域

19. 主要绩效指标

20. 物料审核小组

Ⅱ. Sentences Interpretation

1. 仅仅了解一些质量管理方法而不知道如何实践没有任何好处。

2. 监理部按合同对施工安全、质量、进度、成本等进行管理;如无监理部,则由 MSC-PE 负责项目管理。

3. 桩基工程结束,监理公司按照有关法律法规组织桩基工程验收,取得桩基工程验收合格文件。

4. 质量控制必须由承包商实施,而质量保证则由与所有者有约定的独立的第三方机构来执行。

5. 外区混凝土的性能大大受固化作用影响,因为就是这一区域的混凝土易受湿气、碳酸化和磨损的影响。

Ⅲ. Dialogue Interpretation

【场景:负责人 A 正在和工人 B 在工地上就施工安全话题展开对话。】

A：大臂侧面写的什么？

B："Danger! keep away from under boom!"

A：可我不得不给机器拴绳啊？

B：Leave there as soon as you fixed it.

A：但我必须导入地脚螺栓呀？

B：You have to take a risk then.

A：我认为规定条文没有用。

B：No, it isn't. It tells you to keep safety in mind all the time.

A：对，我同意。"安全第一。"

B：Have you ever seen things dropping down from hoisted height?

A：没，从来没有。

B：That's what used to happen.

A：真的吗？

B：Of course!

A：谢谢你，我一定牢记。

IV. Passage Interpretation

There are many factors that affect the quality of engineering projects, which can be summarized into six main aspects, namely man, material, machine, method, measurement, and environment, which are commonly known as 5MIE factors, and the strict control and comprehensive coordination and management of these six factors is the key to ensuring the quality of the project.

People refer to the decision-makers, managers and operators involved in the construction of engineering projects, and their quality and ability will affect the quality level of engineering projects. When the project manager carries out quality management, the human factor should be considered and controlled from the aspects of the implementer's quality, theoretical and technical level, physiological condition, psychological behavior, etc.

Machinery is an important material basis for the implementation of mechanization of the project, the project manager should comprehensively consider the conditions and characteristics of the construction of the project, formulate a mechanized construction plan, so that the mechanical equipment and construction organization plan are organically linked, and effectively improve the comprehensive benefits of the project.

The method is an important factor to ensure the stable improvement of project quality. The project manager should comprehensively analyze and consider the process, technology, operation, organization, procedure and other aspects according to the actual situation of the project, and strive to improve the project quality, speed up the progress of the project and reduce the project cost.

Measurement refers to whether the measurement method adopted in the construction of the project is standard and correct. The project manager should pay attention to the accuracy required for the measurement task, prepare the corresponding measurement tools, regularly inspect and calibrate the measurement tools and test equipment, and formulate the necessary calibration procedures to reduce the errors generated in the measurement process.

Environment includes engineering and technical environment, engineering operation environment, engineering management environment and surrounding environment and many other environmental factors. These environmental factors are complex and changeable, and the project manager should strengthen environmental management and take necessary measures to control the impact of the environment on quality according to the conditions and characteristics of the project.

第九章　工程项目管理之成本管理

工程项目成本管理是企业一项重要的基础管理,是指施工企业结合本行业的特点,以施工过程中直接耗费为原则,以货币为主要计量单位,对项目从开工到竣工所发生的各项收支进行全面系统的管理,以实现项目施工成本最优化目的的过程。它包括落实项目施工责任成本,制定成本计划,分解成本指标,进行成本控制、成本核算、成本考核和成本监督的过程。其重要性体现在可降低成本并提高竞争力,提高资源使用效率,提高决策准确性以及实现企业经济目标。

第一节　背景知识

工程成本管理是指在工程项目的全过程中,对工程成本进行有效的计划、控制和管理,以实现工程项目的经济效益最大化。一般来说,工程项目的成本管理分为三个阶段,首先是项目施工前期的成本管理,其次是施工期间的成本管理,最后是竣工检验的成本管理。其简单介绍如下:

一、施工前期的成本管理

在施工前期的准备阶段,必须对施工顺序、施工方法、机械选择、作业组织形式和技术措施进行详细的研究和分析,确定经济、合理、可行的施工方案。确定方案后,应根据成本控制目标,根据项目结构分解或项目单元包含的实际工作量,根据消费标准或技术措施等优化施工方案,制定成本控制计划。

二、施工期间的成本管理

论证项目成本控制目标,综合考虑项目整体的统一协调性,不仅要满足项目成本控制目标的要求,还要实现预定的质量和进度目标,实现成本目标、质量目标和进度目标的平衡和优化。同时,根据施工进度,比较实际产生的成本和相应的计划目标,计算成本差异,分析偏差产生的原因,采取有效的控制措施。

三、竣工检验的成本管理

项目完成后,立即申请项目竣工检验,工程顺利交付使用,立即处理相关结算资料,保证结算资料的真实性和完整性。在工程所需的保修期内,必须明确保修的主要负责人,确定保修期的费用管理。

四、成本管理的主要措施

(一)加强施工成本管理

施工前期,应迅速组织设备进入现场,根据施工任务分布情况,在保证合理利用资源和合同工期的前提下,兼顾经济、效率等原则,迅速完成作业区划分,引进具有丰富施工经验的技术人员、管理人员和操作人员完成后,结合施工图纸,根据中标合同价格和合同文件等,收集、分析工程项目所在地的供给条件和市场行情等,制定成本指导价格,确定成本管理的目标。在施工过程中,首先要注意目标责任成本的实施,标志目标责任成本,经成本管理组一致通过、预算经营处审查通过后,由项目经理提出指标。

要熟悉施工组织设计和图纸,结合施工现场情况,了解项目可能产生的利润和影响。施工过程中的直接成本管理,通常有三个部分:人工成本控制、材料成本控制和现场管理成本控制。人工成本的控制需要严格筛选,选择有多年合作经验的好团队,全面推进单价承包工资制度,直接将工作完成情况与收入联系起来,严格划分工人的工作内容和范围,提出定额指标,统一实施合同化的管理模式,降低使用者的消耗。材料费用的控制具有较大的潜力,一般来讲,材料费占据工程总费用的60%～70%,要根据工程的实际情况,制定合理的材料费用控制措施。

(二)规范成本管理

施工的成本管理涉及整个工程中全体参与建设的人员,与每个人的利益息息相关。在实际操作过程中,要采取多种方式进行分阶段、分批次的成本管理培训,树立起全员参与成本管理、人人有责任的管理意识,增强每个员工的成本管理责任心。成本管理需要依靠完善的组织计划以及严密的制度进行规范,做到有据可依、有章可循,避免工作失误,通过合理的、量化的措施来达到降低消耗、降低成本、减少浪费的目标。

(三)做好监控和分析

要做好成本分析工作,及时发现并且纠正在执行过程中出现的偏差,建立起项目成本管理机构,有计划地做好成本原始数据的搜集和整理工作,正确核算各个阶段以及各个工程的成本,根据预算进行考核,分析实际成本与预算成本之间的差异,找到产生这一差距的原因,及时反馈给相关的技术部门,采取积极有效的措施纠正偏差,防止对工程产生不利的影响。

工程成本管理对于工程项目的成功至关重要。首先,工程成本管理可以帮助企业在竞争激烈的市场中,通过降低成本,提高竞争力。其次,工程成本管理可以帮助企业提高资源的使用效率,避免资源的浪费。再次,工程成本管理可以帮助企业提高决策的准确性,避免因成本预测不准确而导致决策失误。最后,工程成本管理可以帮助企业提高工程项目的经济效益,实现企业的经济目标。

第二节　相关词语与表达

序号	中文	英文
1	标准成本	standard cost
2	实际成本	actual cost
3	标准成本差异	standard cost variance
4	材料成本差异	material cost variance
5	人工成本差异	labor cost variance
6	制造费用差异	manufacturing cost variance
7	成本效益分析	cost-benefit analysis
8	直接成本	direct cost
9	间接成本	indirect cost
10	可变成本	variable cost
11	不可变成本	invariable cost
12	预算成本	budgeted cost
13	投资估算	investment estimation
14	建设工程造价咨询	construction project cost consultation
15	成本结构分析	cost structure analysis
16	预算	budget

续表

序号	中文	英文
17	成本控制	cost control
18	成本预算	cost estimate
19	成本超支	cost overrun
20	成本节约	cost saving
21	投资回报率	return on investment（ROI）
22	资本支出	capital expenditures（CAPEX）
23	运营费用	operating expenses（OPEX）
24	成本收益分析	cost-benefit analysis
25	生命周期成本	life-cycle cost（LCC）
26	挣值管理	earned value management（EVM）
27	沉没成本	sunk cost
28	现值	present value（PV）
29	净现值	net present value（NPV）
30	折现现金流量	discounted cash-flow（DCF）
31	内部收益率	internal rate of return（IRR）
32	折现因素	discount factor（DF）
33	投资回报率	return on investment（ROI）
34	收益成本比率	benefit cost ratio（BCR）
35	费用绩效指数	cost performance index（CPI）
36	计划价值	planned value（PV）
37	工程造价	construction cost
38	工程计价	project valuation
39	建设项目总投资	total investment of construction project
40	建设投资	construction investment
41	设备购置费	original equipment cost
42	建筑安装工程费	construction and installation engineering cost
43	工程建设其他费用	other expenses of engineering construction
44	预备费	reserve fund
45	基本预备费	the basic reserve fund
46	价差预备费	reserve for price difference

续表

序号	中文	英文
47	建设期利息	interest incurred during construction
48	不可预见费	contingency sum
49	静态投资	static investment
50	动态投资	dynamic investment
51	人工费	labor cost
52	材料费	material fee
53	机械费	machinery fee
54	企业管理费	enterprise management fee
55	直接费	direct fee
56	间接费	indirect fee
57	直接工程费	direct engineering fee
58	措施费	measures fee
59	安全文明施工费	safety civilized construction measures
60	夜间施工增加费	night construction increase fee
61	二次搬运费	secondary handling fee
62	冬雨季施工增加费	winter and rainy season construction increase fee
63	施工排水费	construction drainage fee
64	施工降水费	construction precipitation fee
65	已完工程保护费	completed engineering protection fee
66	规费	stipulated fee
67	税金	taxation
68	销项税额	output VAT
69	建设管理费	construction management fee
70	建设用地费	construction land fee
71	可行性研究费	feasibility study fee
72	研究试验费	research and test fee
73	勘察设计费	investigation and design fee
74	环境影响评价费	environmental impact assessment fee
75	劳动安全卫生评价费	labor safety and health evaluation fee
76	建设项目场地准备费	site preparation fee for construction project

续表

序号	中文	英文
77	建设单位临时设施费	temporary facilities fee for construction unit
78	引进技术和设备的其他费用	other fees for imported technology and equipment
79	工程保险费	engineering insurance premium
80	联合试运转费	joint commissioning fee
81	特殊设备安全监督检验费	special equipment safety supervision and inspection fee
82	市政公用设施费	municipal public facilities fee
83	无形资产费用	intangible assets fee
84	其他资产费用	other assets fee
85	生产准备费	production preparation fee
86	人工单价	artificial price
87	设备原价	equipment original price
88	到岸价	cost, insurance and freight（CIF）
89	进口从属费	import subsidiary fee
90	银行财务费	bank finance fee
91	外贸手续费	foreign trade handling fee
92	进口关税	import tariff
93	消费税	consumption tax
94	进口环节增值税	import value-added tax
95	车辆购置税	vehicle purchase tax
96	设备运杂费	equipment transportation and miscellaneous fee
97	材料单价	material unit price
98	材料原价	material original price
99	材料运杂费	material transportation and miscellaneous fee
100	采购及保管费	purchase and storage fee
101	检验试验费	inspection and test fee
102	机械台班单价	machine shift unit price
103	工程量清单	bill of quantities
104	投标报价细目	breakdown of tender prices
105	招标工程量清单	bidding bill of quantities
106	已标价工程量清单	priced bill of quantities

续表

序号	中文	英文
107	暂扣款,保留金	retention money
108	综合单价	comprehensive unit price
109	机械使用费	machinery use fee
110	仪表使用费	instrument use fee
111	企业管理费	enterprise management fee
112	措施项目	measures project
113	分部分项工程费用	partial project cost
114	措施项目费	measures project cost
115	其他项目费	other project cost
116	暂定费	provisional sum
117	工程进度款	progress payment
118	计日工费	daily work fee
119	总承包服务费	general contracting service fee
120	工程价款调整	project price adjustment
121	单价项目	unit price project
122	总价项目	total price project
123	工程量偏差	engineering quantity deviation
124	索赔	claim
125	现场签证	on-site visa
126	企业定额	enterprise quota
127	违约罚金	liquidated damages
128	文明生产费	civilized production fee
129	文明施工费	civilized construction measure
130	环境保护费	environmental protection fee
131	安全生产费	safety production fee
132	工程定额	project quota
133	预算定额	budget quota
134	概算指标	estimate index
135	劳动定额	labor quota
136	施工定额	construction quota

续表

序号	中文	英文
137	投资估算指标	investment estimate index
138	工期定额	time limit quota
139	定额基价	quota base price
140	人工消耗量	labor consumption
141	材料消耗量	material consumption
142	机械台班消耗量	machine-team consumption
143	工程计价信息	project valuation information
144	工程造价指数	project cost index
145	专业造价工程师	professional cost engineer
146	造价工程师	quantity surveyor
147	持证概预算员	certified budget estimator
148	发包人	employer
149	承包人	contractor
150	分包人	subcontractor
151	工程造价咨询人	project cost consultant
152	招标代理人	bidding agent
153	承包人项目经理	contractor's project manager
154	监理人	supervisor
155	总监理工程师(总监)	chief supervision engineer (director)
156	总造价工程师	chief cost engineer
157	全过程造价管理	whole process cost management
158	全方位造价管理	all-round cost management
159	全要素造价管理	total factor cost management
160	全生命周期造价管理	whole life cycle cost management
161	工程造价控制	project cost control
162	工程造价确定	project cost determination
163	投资估算	investment estimation
164	设计概算	design estimate
165	施工图预算	construction drawing budget
166	招标控制价	bidding control price

续表

序号	中文	英文
167	财务报表	financial statement
168	合同价	contract price
169	工程结算	project settlement
170	工程造价纠纷鉴定	appraisal of engineering cost disputes
171	工程造价成果质量鉴定	quality appraisal of engineering cost achievement
172	工程项目风险管理	risk management of engineering project
173	方案比选	scheme comparison
174	限额设计	quota design
175	优化设计	optimization design
176	深化设计	deepening design
177	价值工程	value engineering
178	投资估算	investment estimation
179	生产能力指数法	productivity index method
180	系数估算法	coefficient estimation method
181	比例估算法	proportion estimation method
182	指标估算法	index estimation method
183	综合估算指标	comprehensive estimation index
184	单项工程估算指标	single project estimation index
185	单位工程估算指标	unit project estimation index
186	通融付款	ex gratia payment
187	流动资金的扩大指标估算法	expanded index estimation method of liquidity
188	分年度投资计划	annual investment plan
189	经营成本	operating cost
190	设计概算	design estimate
191	修正设计概算	revised design estimate
192	总概算	total estimate
193	单项工程综合概算	comprehensive budget estimate of single project
194	单位工程概算	unit project budget estimate
195	概算定额法	budget estimate quota method
196	概算指标法	budget estimate index method

第三节　对话口译

Dialogue 1：成本管理说明（音频：9.3.1）

Word Tips

cost management	成本管理
fixed costs	固定成本
variable costs	可变成本
cost analysis	成本分析
cost accounting software	成本核算软件
generate reports	生成报告
cost-benefit analysis	成本效益分析
profitability	收益性
revenue	收入

【场景：Tom and Li Ling work together in the company. When they are chatting together, Tom asks Li Ling some questions about cost management.

汤姆和李玲在公司里共事，在一起聊天时，汤姆问了李玲一些关于成本管理的问题。】

李玲：嘿，汤姆，你了解成本管理吗？

Tom: Yeah, sure. Cost management is the practice of identifying, analyzing, and controlling costs in order to improve business performance.

李玲：那很有趣。一般公司是如何进行成本管理的？

Tom: Well, there are different strategies that companies use to manage their costs. For example, a company might focus on reducing its fixed costs, such as rent and salaries, or it might try to minimize its variable costs, such as materials and supplies.

李玲：我明白了。那成本分析呢？你能给我解释一下吗？

Tom: Sure. Cost analysis involves scrutinizing all the costs associated with a particular project or process, so that you can identify which costs are necessary and which ones are unnecessary. Once you have done that, you can start to explore ways to reduce costs wherever possible.

李玲:有道理。公司有没有管理成本的工具?

Tom: Yes, there are lots of different tools and techniques. For example, many companies use cost accounting software to keep track of their expenses and generate reports. They might also use cost-benefit analysis to determine whether a particular investment is worth making.

李玲:好吧。成本管理如何影响公司的盈利?

Tom: Well, effective cost management can help a company improve its profitability, because it allows the company to reduce its expenses and increase its revenue. It can also help the company stay competitive, because it allows it to offer its products or services at a lower price than its competitors.

李玲:好的,汤姆,谢谢你给我讲解。

Tom: No problem, Li Ling. I'm always happy to help.

Dialogue 2:项目成本管理会议（音频:9.3.2）

Word Tips

budget table	预算表
research and development stage	研发阶段
implementation phase	执行阶段
technician	技术员
procurement cost	采购成本
debugging and optimization work	调试和优化工作
final delivery	最终交付
support and maintenance costs	支持和维护成本
after-sales service	售后服务
monitoring cost	监控成本

【场景:The project manager (Wang Lin—responsible for the planning and implementation of the project), the financial specialist (Mary—responsible for the tracking and reporting of the project cost) and the engineer (Jerry—responsible for the technical implementation of the project) are discussing a new project. They need

— 173 —

to effectively manage the project cost to ensure that the project is carried out as planned and completed within the budget. 】

项目经理（王林——负责项目的计划和执行），财务专员（Mary——负责项目成本的跟踪和报告）和工程师（Jerry——负责项目的技术实施）正在讨论一个新的工程项目。他们需要有效地管理项目成本以确保项目按计划进行并在预算内完成。

王林:大家好,感谢你们参加这次项目成本管理会议。我们需要确保我们的项目能够按照计划进行,并在预算范围内。

Jerry: Yes, the technical requirements of this project are very high, and we need to ensure that the cost is controlled without sacrificing quality.

Marry: I have prepared the budget table, and we can discuss it from each stage.

王林:好的,让我们从项目的起始阶段开始。Jerry,你认为在项目初期有哪些成本可能会增加?

Jerry: It is mainly the experimental and testing costs in the research and development stage, and we may need to buy some special equipment.

Mary: I'll put these into the budget table. Besides, are there any other potential costs?

Jerry: There are also training costs, because we may need team members to master some new technologies.

王林:好,这些都很重要。现在让我们看看项目执行阶段。Jerry,你对执行阶段的成本有什么看法?

Jerry: The first is the labor cost, including the salaries of engineers and technicians. In addition, the procurement cost of materials and equipment is also a consideration.

Mary: I'll put these in the table in detail. Are there any other factors that may affect the cost of the implementation phase?

Jerry: There may be some unknown technical challenges, resulting in additional R&D and testing costs. We need to leave some flexibility in the budget to deal with these situations.

王林:明白了。最后,让我们看看项目完成阶段。有哪些成本可能会在这个阶段产生?

Jerry: There may be some extra debugging and optimization work to ensure that the final delivery of the project meets the customer's expectations.

Mary: Okay, is there anything else to consider?

王林:比如,项目交付后的支持和维护成本。我们需要确保客户在使用我们的产品时能够得到及时的支持。

Jerry: Yes, this is also an important aspect. We need to reserve some resources for after-sales service in the budget.

王林:非常感谢大家的反馈。Mary,你能在表格中整理一下我们讨论的所有成本吗?

Mary: Of course, I will reflect all the cost forecasts in the table and make sure that we have enough budget to cover the needs at all stages.

王林:非常好,谢谢大家的参与。我们会持续监控成本,并在需要时做出调整,以确保项目的成功完成。

第四节　篇章口译

Passage 1:成本管理工作总结（音频:9.4.1）

Word Tips

cost management	成本管理
launch	发行
infrastructure construction projects	基建项目
anniversary	周年纪念日
cost-effectiveness	成本效益
cost accounting and monitoring mechanism	成本核算和监控机制
transparency	透明度
cost overruns	成本超支
project implementation	项目实施

post   岗位

【场景：A project manager published a summary of the cost management work on the overseas infrastructure construction projects carried out by the company.

某项目经理就本公司所进行的海外基础设施建设项目发表关于成本管理工作的总结。】

Your excellencies and dear colleagues:

Hello everyone! I am very glad to share with you our cost management in overseas infrastructure construction project at this special moment. Today is the first anniversary of the launch of our project. Looking back on the past year, we not only made a lot of progress, but also faced some challenges, but through the joint efforts of the team, we have made remarkable achievements.

To begin with, I want to emphasize the importance of cost management in our project. In the context of fierce competition in the international market, we must pay more attention to cost-effectiveness and ensure that our projects can be completed efficiently and remain competitive. Cost management is not only a financial issue, but also the lifeline of the project, which is related to our company's reputation and competitive advantage in the international market.

In the past year, we have been constantly striving to improve the level of cost management. We have taken a series of measures, through which we have successfully reduced some unnecessary expenses and improved the overall benefit of the project.

First of all, I want to emphasize our strict management of costs. We clearly realize that cost control is not only related to finance, but also an important factor to ensure the sustainability and success of the project. We are committed to accurately controlling every expenditure without affecting the quality and progress of the project. This means that we not only pay attention to the current expenditure, but also pay attention to the return and benefit of long-term investment.

Secondly, we have taken a series of innovative cost management measures. From the beginning of the project, we have established a strict cost accounting and monitoring mechanism to ensure the transparency and efficiency of the use of funds.

We use advanced technology and data analysis tools to identify potential cost risks in time and take measures to solve them quickly to avoid cost overruns.

Thirdly, we also attach importance to teamwork and communication. Cost management is not only the responsibility of the financial team, but also the common mission of all project participants. We encourage communication and collaboration among team members to ensure that everyone can realize the importance of cost control and actively participate in the practice of cost management.

However, we should also be soberly aware that cost management is a process of continuous improvement, which requires us to constantly sum up experience, find problems and make timely adjustments. In the future project implementation, we will further strengthen cooperation with suppliers, deepen cost accounting and analysis, and find more saving space. At the same time, we will also pay attention to the training of team members, improve the whole team's awareness of cost management, so that everyone can play an active role in their posts.

Finally, I want to thank every team member for their hard work in the past year. It is because of everyone's efforts that we can achieve today's results. I hope that in the future, we can, as always, unite and cooperate to overcome all kinds of difficulties and work hard for the success of the project.

Thank you!

Passage 2:有关项目成本管理的建议（音频:9.4.2）

Word Tips

投标	bid/tender
签约	sign a contract
亏损额度	loss limit
报价	quote
招标文件	bidding document
条款	provision
资信	creditworthiness
履约能力	contractual capacity

标书费	tender fee
中标率	successful rate
合同交底	contract disclosure
盈利点	profit point
项目亏损	project loss
设计图纸	design drawing
劳务施工分包商	labor construction subcontractor
劳务分包价	labor subcontracting price
现场零星用工	sporadic employment on site
包干	lump sum
买价控制	purchase price control
大宗材料	bulk materials
竞价	competitive bidding
周转材料	turnover material
承包机制	contracting mechanism
项目工薪	project salary
资格审查	qualification review
竞争机制	competition mechanism
预付款	advance payment
保函	letter of guarantee
质保金	quality guarantee deposit
办理分包结算	handle subcontracting settlement
竣工验收阶段	completion acceptance stage
交付	delivery

【场景：A project manager put forward some methods and suggestions on project cost management based on the actual project cost management of the enterprise and his own work practice.

某项目经理以企业的实际项目成本管理情况为基础,结合自身工作实践,对项目成本管理提出了一些方法和建议。】

大家好,我将结合自身的工作实践和我们的项目成本管理现状,并结合本企业的实际对项目成本管理进行探讨,提出加强项目成本控制的方法和建议。

一是投标、签约阶段的成本控制。

随着市场经济的发展,施工企业正处于"找米下锅"的紧张状态,忙于找信息,忙于搞投标,忙于找关系。为了中标,施工企业把标价越压越低。有的工程项目,管理稍一放松,就要发生亏损,有的项目亏损额度较大。因此,做好标前成本预测,科学合理地计算投标价格及投标决策,显得尤为重要。在投标报价时要认真确认招标文件所涉及的每一项经济条款,了解业主的资信及履约能力,确有把握再做标。在做完后、报出前,要组织有关专业人员进行评审和论证,在此基础上,再报企业领导最后决策。

为做好标前成本预测,企业要根据市场行情,不断收集、整理、完善符合本企业实际的内部价格体系,为快速准确地预测标前成本提供有力保证。同时,投标也会产生多种费用,包括标书费、差旅费、咨询费、办公费、招待费等。因此,提高中标率、节约投标费用,也成为降低成本开支的一项重要内容。投标费用要与和中标价相关联的指标挂钩,实施总额控制,规范开支范围和数额,并由一名企业领导专门负责招投标工作及管理。中标后企业在合同签约时要据理力争,尤其是目前的开发商,将不利于施工企业的合同条件在投标阶段已列入招标文件,并且施工企业在投标时已确认招标文件,要想改变非常困难。但是,也要充分利用签约的机会,尽量与业主协商相关不利的条款,尽可能做到公平、合理,力争将风险降至最小后再与业主签约。

签约后,公司要认真组织机关及项目部相关部门的有关人员进行合同交底,通过不同形式的交底,使项目部的相关管理人员明确本施工合同的全部相关条款和内容,为下一步扩大项目管理的盈利点、减少项目亏损打下了基础。

二是施工准备阶段的成本控制。

根据设计图纸和有关技术资料,结合本工程的实际,大力开展新技术、新工艺的应用,对施工方法、施工程序、作业组织形式、机械设备选型、技术组织措施等进行认真的研究分析,制定科学先进、经济合理的施工方案。

三是施工过程中的成本控制,其中包括四个方面的内容。

1.要控制好人工、材料、机械及现场管理费

人工费成本主要从分包劳务费方面进行控制。要根据工程的实际,以招投标的形式选定劳务施工分包商,确定劳务分包价。辅料、小型机具及现场零星

用工,包括垃圾清运等金额不大又不好控制的方面,可采取费用包干形式一次包死。

材料费成本的控制包括材料用量控制和材料价格控制两方面。材料用量要坚持限额领料制度,并推广使用降低料耗的新技术、新工艺、新材料,坚持余料回收,降低料耗水平,降低堆放、仓储损耗。材料购入时要实行买价控制,大宗材料应引入市场竞争机制在保质保量的前提下竞价购料。包括预拌砼的使用,价格要通过竞价招标,数量要严格按施工图纸控制结算。进入施工现场的其他所有材料要严格实行收料及保管制度,确保质量、确保应有的数量。同时要考虑资金的时间价值,减少资金占用,降低存货成本。不好控制及管理的辅助性材料,经过测算后,可采取费用包干的形式一次性含在劳务分包价中,由劳务分包商包干使用。

其他费用,主要包括工程水电费、垃圾清运等直接费用。对此,要本着节约原则,推行承包机制,对现场各类分包商实行收费管理,减少开支、节约成本。

现场管理费,亦称项目间接费,包括临时设施费和现场经费,该费用中开支较大的主要是临时设施、项目工薪、交通费和业务招待费等。为把该费用开支降下来,应该做到:制定开支计划,精减管理人员,实施总额控制;严格控制招待费用,对各项费用按费用性质,落实责任部门和人员;对于特殊性开支和较大数额开支,要开会研究,报企业领导审批。临时设施要根据工程及现场的实际,本着节约、科学合理,尽可能实用,并且能多次周转使用的原则进行布置和建设。

2. 要加强施工组织,合理使用资源,降低工期成本

在合理的工期下,项目成本支出较低。相较于合理工期,工期提前或拖后都意味着工程成本提高。因此,在安排工期时,项目经理应考虑工期与成本的辩证统一关系,组织均衡施工,以求在合理使用资源的前提下,保证工期,降低成本(若因建设方原因,工期要提前或推迟,则必须与建设方办理洽商及索赔)。

3. 加强安全、质量管理,控制安全和质量成本

安全和质量成本包括两个方面,即控制成本和事故成本。要实行安全生产、文明施工,提高产品质量,适当地控制成本是必要的,需要降低的是事故成本,即发生事故时项目的损失。事故成本是工程无安全、质量事故时就会消失的成本,我们必须加强安全、质量管理,使安全、质量的事故成本降至最低。

4.加强分包管理,控制分包成本

对于分包工程(劳务及各类专业分包),除了严格对分包队伍的资格进行审查,还要充分引入市场经济的竞争机制,实施招标竞价,科学、合理地确定分包工程价格。采取严格的管理手段,包括合同的签订,预付款和工程款的支付,保函、质保金的扣留,现场要更加严格地实施全方位的监控、管理。完工后及时办理分包结算,锁定分包成本。

四是竣工验收阶段的成本控制。

从现实情况看,很多工程从开工到竣工扫尾阶段,就把主要技术力量抽调到其他在建工程,以致扫尾工作拖拖拉拉,战线拉得很长,机械、设备无法转移,成本费用照常产生,使已取得的经济效益逐步流失。因此,要精心安排,力争把竣工扫尾时间缩短到最低限度,以降低竣工阶段的成本支出。特别要重视竣工验收工作,在验收以前,要准备好验收所需要的种种书面资料,送甲方备查;验收中甲方提出的意见,应根据设计要求和合同内容认真处理,确保顺利交付。

第五节　口译技能——顺句驱动

一、顺句驱动及其效用

顺句驱动,顾名思义就是顺着翻译。它具体是指在口译活动中,译员不受译入语的结构限制,听到什么就翻译什么,在最短的时间内,利用断句、填补、重复的方法,将源语言转化为符合目的语习惯的表达。

口译员的思索时间较短,而顺句驱动的应用便可以大大缩短翻译时间。译员在翻译的时候,不用等说话人将整句话说完再调整语序,这样大大缩短了译员的思索时间,从而提高了口译速度,同时也减轻了译员的短时记忆压力,可以优化分配精力,进而提高翻译质量。例如,英语语言中的定语从句一般后置,放于修饰语的后面;笔译的时候,译者会浏览整句话,具有充分的时间思索,译为汉语时,定语从句是移到修饰语之前还是另起一句话进行翻译。而口译时译员没有时间思考,此时便可以直接采用顺句驱动的方法,即顺着翻译。

二、顺句驱动指导下的具体翻译方法

顺句驱动为口译的首要原则。在顺句驱动原则的指导下,口译过程中还会用到断句、重复、增补、词性转换等具体方法。

（一）断句法

在口译过程中，要根据讲话者的讲话信息合理断句，方便整合信息。断句时划分的意群，根据需要可长可短。它可以是一个词语、一个短语，或者是一句话。断句之后，再按顺序翻译。

例1：That year, we produced a car /that we independently developed.

翻译：那一年，我们生产了一辆轿车，是我们自主研发的。

在这句话中，口译时就要遵循顺句驱动的原则，将定语从句断开，拆成两句话翻译。

（二）重复法

口译中，在顺句驱动的翻译策略下，为了保证语义完整、句子流畅，可能需要将上述已经提到的词语再次重复一遍。

例2：This country has taken giant steps forwards on the path of democracy, equality and national reconciliation.

翻译：该国已经迈出了大步，在民主、平等和民族和解的道路上迈出了大步。

该句在翻译的时候，就重复了"迈出了大步"，若不重复此部分的话，很难让听者理解句子的含义。

（三）增补法

例3：The General Assembly recognized the importance of international cooperation and of studying their underlying causes with a view to finding just and peaceful solutions.

翻译：联合国大会认识到，重要的是进行国际合作；<u>同样重要的是</u>研究其根本原因，以便找出公正、和平的解决方法。

译文中下划线部分则是翻译时增补的连接上下文的词语。口译中这样的例子有很多，在翻译时适当地增添词语，以保证译文表达的流畅性和内容的完整性。如果没有下划线标注的成分，译文则不能成为一个语法正确、语意完整的句子。

（四）词性转换法

例4：We appreciate the interest that government has shown regarding the problems of rural areas.

翻译:我们赞赏政府关注关于农村地区所遇到的问题。

因为译员先听到的是"the interest",然后听到从句,根据顺句驱动原则,就应该先翻译"the interest",所以将它翻译为"关注",名词转换成了动词。当然,类似的还有介词转化为动词、动词转化为名词等多种转换形式。

总之,无论是同声传译还是交替传译,留给译员的时间都极其短暂,而顺句驱动恰好能帮助译员在最短的时间内,顺利完成口译任务。本节中提出的具体翻译方法,是以顺句驱动的应用为前提,常常应用在口译当中。

第六节　课后练习

(音频:9.6)

Ⅰ. Technical Terms Interpretation

1. 误期赔偿费

2. 工程进度款

3. 形式发票

4. 投资偏差

5. 量化索赔

6. 工程经济纠纷

7. 竣工决算价

8. 工程全险

9. 现金流

10. 质量保修金

11. earned value(EV)

12. actual cost(AC)

13. budget at completion(BAC)

14. variance at completion(VAC)

15. cost variance(CV)

16. schedule variance(SV)

17. cost performance index(CPI)

18. interim payment certificate

19. workmen compensation insurance

20. interest-free loan

II. Sentences Interpretation

1. 我看了一下明细,你们预算的金额太高了,能不能再压缩一下成本呢?

2. 合理选择机械设备对成本管理有十分重要的意义,尤其是高层建筑。

3. 项目部应从减少因安排不当而闲置的设备、提高设备利用率、避免不正确使用造成机械设备停置等方面控制施工机械使用费。

4. 现场经费应精减,加强施工质量管理,促进管理水平不断提高,减少管理费用。

5. 技术与经济是相互依存的,技术的提高或新技术的采用,必然大幅度提高劳动效率和节省材料,从而节约成本。

6. 譬如,天棚面平整度如果很高,可以不用抹灰,而通过刮泥子的方法一样能达到使用要求,还能加大使用空间,节省费用。

7. 根据项目部制定的考核制度,对成本管理相关责任人员进行考核,考核的重点是完成工作量、材料、人工费及机械使用费四大指标。

8. 为了对施工分包费进行有效控制,项目部主要做好分包工程询价、订立平等互利的分包合同,建立稳定的分包关系网络。

9. 应主要通过掌握市场信息,应用招标和询价等方式控制材料、设备的采购价格。

10. 技术报价书中有很多资料,包括工艺流程、工艺说明、生产能力、产品特性等。

III. Dialogue Interpretation

【场景:The supplier cost of a company is too high, which leads to the company's inability to improve its market competitiveness in the fierce market competition.

某家公司的供应商成本过高,在市场竞争激烈的情况下,导致公司无法提高市场竞争力。】

采购经理 A:Recently, our cost is out of control, and the price of suppliers is getting higher and higher, which makes us more and more stressed.

总经理 B:这肯定不行的,我们要想办法控制成本。

采购经理 A:Yes, I think we can renegotiate the contract with the supplier and seek more favorable terms. This may include price renegotiation, optimization of delivery time, quantity discount, etc.

总经理 B:好办法! 不过我们也得多考虑其他对策,因为谈判可能不会达成一致。

采购经理 A:Then we can replace the original suppliers and find new suppliers, or consider establishing cooperative relations with multiple suppliers to reduce our dependence on a single supplier. In this way, we can negotiate the price better and have an alternative plan when a supplier has problems.

总经理 B:对的,我认为还是要看品质、服务等因素,不一定价格低就能行。我们应该对原有供应商的价格、质量、服务等方面进行评价,并与其他供应商进行比较,选出性价比最高的供应商,并与其合作。同时公司还需与供应商沟通,寻找优惠扣减或其他合作方式以降低成本。

采购经理 A:I think what you said is very reasonable, and then I will make a reasonable plan with our colleagues.

Ⅳ. Passage Interpretation

The cost of construction plan is based on the construction cost control plan specific construction projects, including the specific cost predetermined control objectives, and implementation measures and planning control target, is a guiding document for the construction cost control. The project department should carefully prepare construction organization design before the construction, optimize the construction scheme. Construction scheme mainly includes four aspects: construction method of choice, construction machinery selection, construction sequence arrangement and the flow construction organization. The construction scheme, construction period will be different, the required equipment is different. Therefore, optimization of construction scheme selection is the main way to reduce engineering cost in project.

第十章　工程项目验收与付款

第一节　背景知识

一、工程验收

工程验收是指完成工程审查后,上级部门对工程所做的检查验收工作。它是以批准的设计任务书和设计文件、国家或部门颁发的施工验收规范和质量检验标准为依据,由建设单位、施工单位和项目验收委员会按照一定的程序和手续,在项目建成后,对工程项目进行检验和认证的活动。

对于投资者来说,工程验收是其综合检验投资项目是否实现预期目标、审查工程进度和工程质量的重要步骤。通过这个环节,投资者可以更好地控制工程的质量。这关系到工程项目的经济利益、项目投产以后的运营效果以及销售后的经济效益。对承包商来说,工程接受验收是其接受投资者和权威机构全面检查、按合同履行义务、配合投资者组织试生产、办理工程移交的过程。这样,承包商也能就此总结经验,提高施工管理水平。

工程项目的竣工验收是施工全过程的最后一道程序,也是工程项目管理的最后一项工作。它是建设投资成果转入生产或使用的标志,也是全面考核投资效益、检验设计和施工质量的重要环节,工程验收直接关系到项目日后的正常运营以及投资效益的实现。

（一）工程验收的准备工作

工程验收准备工作分为两个方面:一方面是承包商应做的工作,包括做好工程项目的收尾工作、编制工程项目竣工验收报告、整理汇集各种经济与技术资料、拟定验收条件及验收依据等验收的必备技术资料;另一方面是验收方应做的工作,包括组成项目验收工作组或验收委员会、项目材料验收、现场初步验收、正式验收、签发项目验收合格文件以及办理固定资产形成和增列手续。

（二）初步交工验收

承包工程竣工后,双方及有关管理部门如监理公司、设计单位等共同组成验收小组,对工程进行全面检查。经检查合格,由业主签发初步交工验收证书给承包商。从业主签发初步交工验收证书之日起,有一年的保修期。在此期间内,工程的质量问题将由承包商出资解决。

(三)最终交工验收

在一年保修期满后,业主将签发最终交工验收证书。事前,承包商必须准备好下述文件:财务账目审计合格单、国家审计署签认单、完税手续、税务局签认单、海关清白证书、海关签认单、上报统计资料、国家统计局签认单、法院和警察局出具的刑事犯罪案结案情况、供水局出具的施工用水财物两清证明、供电局出具的施工用电财物两清证明、城建局出具的施工用地财物两清证明、实验中心出具的实验中心财物两清证明、保险公司出具的保险公司财物两清证明。承包商准备好以上文件,经业主检查认定完整合格,即签订最终交工验收证书。承包商取得最终交工验收证书,才能据此在银行办理5%的工程保留金清退和工程余款清付手续。

二、工程付款

付款是国际承包工程中承包商最为关心而又最为棘手的问题。承包工程的风险或者利润,最终都要在付款中表现出来。因此,在签订工程合同和实施工程的过程中,承包商在与付款有关的各种问题上需要花费相当多的精力。有关付款的问题可以概括为以下三个方面:价格问题,即承包该项工程,究竟可以得到多少工程款;货币问题,即业主将用何种货币支付;支付方式问题,即业主将在什么时间用什么方法向承包商付款。

(一)价格问题

国际承包工程的合同计价方式有许多种,总结起来可以划分为三大类型:固定总价——合同的总价一般来说是不允许调整的;固定单价——根据单位工程量的固定价格和实际完成的工程量计算实际总价;成本加酬金——成本费用按实际花费全部由业主支付,并给承包商一定数额或固定百分比的管理费和利润。对于承包商来讲,不同的计价方式,其风险也各不相同。

(二)货币问题

由于国际货币市场上难以预料的变化,承包商会碰到很多令人头痛的货币兑换限制、货币汇率浮动、货币支付等问题。在国际承包工程中,承包商不可能只使用一种货币,而是要使用两三种甚至更多种货币。因此,承包商希望得到的是一种可以任意转换成其他货币并且不受货币价值浮动影响的硬通货。一般来说,业主喜欢用当地货币作为支付手段,因为这样做业主可以不承担任何货币价格浮动的风险。由于承包商和业主都不愿意承担货币价值浮动的风险,双方只好协商达成一种妥协的方案,共同承担这类风险。

(三)支付方式

主要考虑支付时间、支付方法和支付保证等问题。

承包商要努力避免使自己成为一个建设项目的实际财务贷款人,除非在签订合同时就已经明确。在支付时间上,对承包商来说,当然是越早得到工程款越好,这样可以减少承包商垫付的周转资金及利息;业主则总是希望延迟付款。国际承包合同惯用的支付大致划分为四段,即预付款、工程进度付款、最终付款和退还保留金。

1. 预付款

用来支付承包商的初期准备费用。支付预付款是公平合理的,因为承包商早期使用的资金相当大。

2. 工程进度付款

一般随着工程按月支付。

3. 最终付款

工程完工后结算付款。一般来说,当工程完工并经现场工程师验收签发合格证明时,意味着工程竣工。这时,承包商有权取得全部工程的合同价款中一切尚未付清的款项。

4. 退还保留金

保留金是业主从每次工程进度款中扣留的一定金额。承包商应当在商签合同时就保留金问题与业主磋商,签订公平合理的协议。

关于付款的保证是极为重要的。通常的办法是要求工程业主提交一份银行开出的信用证。对于延期付款或卖方信贷项目,业主是没有理由拒绝开出信用证的。如果业主是私营公司,或者业主虽然是政府部门,但建设资金属于外国基金,也应当要求业主开出银行的信用证。如果不信任开证银行的资信情况,还可以要求另一家承包商接受的银行对开证行的信用证进行保兑。信用证应当是不可撤销的。承包商要注意信用证是否有时间限制,因为工程延期是十分常见的现象,而且原因可能是多方面的。有限期的信用证一旦过期,银行可以免除付款责任,这对承包商是不利的。

　　即使采用各种预防措施,有时也难以完全避免拖欠付款。承包商只有事先对业主的支付能力和信誉进行充分调查,选择承包那些确有付款保证的项目,这样才能避免较大的风险。

第二节　相关词语与表达

序号	中文	英文
1	验收	acceptance
2	验收方	acceptor
3	验收程序	taking-over procedure
4	移交	transfer
5	过期	expire
6	审计	audit
7	核实	verify
8	采购	procurement
9	清仓销售	closeout
10	扣留	retention
11	实质性竣工	substantial completion
12	工程竣工验收	final acceptance of project
13	质量监督员	quality supervisor
14	专门小组	expert panel
15	技术员工	technical staff
16	移交证书	taking-over certificate
17	项目启动	project initiation
18	验收动态	dynamics of acceptance
19	按计划	on schedule
20	额定的测试结果	specified test result
21	生产模式	production mode
22	书面通知	written notification
23	验收文件	acceptance document

续表

序号	中文	英文
24	质量保证部经理	quality assurance manager
25	主要的出面人员	primary interface
26	实施标准	implementation criteria
27	项目审计	project audit
28	质量和性能要求	quality and functional requirements
29	结清	close out
30	预付款	advance payment
31	预收款	advance receipt
32	预付汇款	advance remittance
33	预付租金	advance rental payment
34	保留金	retention money
35	进度付款	progress payment
36	附加的	additional
37	附加费用	additional charge（additional cost）
38	附加运费	additional freight
39	增加人工投入	additional input of labor
40	额外付款	additional payment
41	附加费率	additional rate
42	附加税	additional tax
43	校正	adjust
44	调整后的价格	adjusted price
45	管理成本	management cost
46	管理费	management expenses
47	已上涨的价格	advanced price
48	投标包干价格分解	breakdown of lump sum bid price
49	单价分析	breakdown of price
50	分项价格	breakdown price
51	营业税	business tax（tax on business）
52	资本收益税	capital gains tax
53	解除合同补偿费	compensation for cancellation of contract

续表

序号	中文	英文
54	损害赔偿	compensation for damage
55	遣散费	compensation for removal
56	附加费用的补偿	compensation of additional cost
57	每日津贴	day allowances
58	计日工	daywork
59	计日工资	daywork rate
60	直接成本	direct cost
61	直接人工费	direct labor cost
62	直接损失	direct loss
63	直接管理费	direct overhead
64	完税前交货价	duty unpaid price
65	完税价值	duty-paying value
66	有效合同价	effective contract price
67	开办费	establishment charges
68	征税	levy
69	工资单	payroll
70	工资支票	payroll check
71	工资表格	payroll form
72	罚金	penal sum
73	罚款	penalty
74	期间成本	period cost
75	个人所得税	personal income tax
76	零用现金	petty cash
77	开办费	preliminary expenses
78	开办项目	preliminary items
79	预付费用	prepaid expenses
80	利润加成	profit mark-up
81	利润额	profit margin
82	利润率	profit rate
83	财产税	property tax

续表

序号	中文	英文
84	临时付款	provisional payment
85	暂定金额	provisional sum
86	退款	refund
87	（尤指律师、裁决员等的）聘请费	retainer fee
88	印花税票	revenue stamp
89	印花税	stamp duty（tax）
90	启动成本	start-up cost
91	增值税	value added tax
92	福利费	welfare benefits
93	福利费用	welfare expenses
94	已兑现支票	cashed check
95	保付支票	certified check
96	拒付通知(声明)	notice（statement）of dishonour
97	过期支票	out-of-date check
98	未兑现支票	outstanding check
99	付讫支票	paid check/paid-up-check/cancelled check
100	预填日期支票	post-dated check（PDC）
101	期票	promissory note
102	退票	protested check
103	拒付支票	rejected check
104	未兑现支票	uncashed check
105	作废的支票	voided check
106	发票日后付款汇票	after date bill
107	汇票	bill of exchange
108	未到期汇票	bill undue
109	定期汇票	date draft
110	定期票	day bill
111	即期汇票	sight bill/demand bill/demand draft

序号	中文	英文
112	到期票据	due bill
113	银行汇票	banker's draft
114	商业汇票	commercial draft
115	光票	clean bill
116	跟单汇票	documentary bill
117	正本汇票	original bill
118	远期汇票	time bill
119	汇票支付日期	maturity of a draft
120	开立30天的期票	draft(s) to be drawn at 30 days sight
121	财务票据	financial instrument
122	拒付通知单	protest jacket
123	拒付通知书	protest note
124	拒付证书	certificate of dishonor
125	形式发票,预开发票	proforma invoice
126	发票	invoice
127	正式发票	formal(official) invoice
128	附有单证的发票	invoice with document attached
129	商业发票	commercial invoice
130	领事发票	consular invoice
131	海关发票	customs invoice
132	厂商发票	manufacturer's invoice
133	临时发票	provisional invoice
134	一式两份	in duplicate
135	财务报表	financial statement
136	财务审计报表	audited financial statement
137	资产负债表	balance sheet
138	决算表	final statement
139	财务报告	financial report
140	财务分析	financial analysis

续表

序号	中文	英文
141	财务建议书	financial proposal
142	财务结算	financial settlement
143	财务账目	financial accounts
144	融资计划	financing plan
145	交货付款	cash against delivery/payment on delivery
146	货到付款	cash on delivery（COD）/payment on arrival
147	装运付款	cash on shipment(COS)
148	现金结算	cash with order（CWO）
149	拖期付款	delayed payment
150	直接支付	direct payment
151	一次性付款	one-off payment
152	现金支付	cash payment
153	现金购买	cash purchase
154	现金结算	cash settlement
155	支票到期日	check due date
156	收款银行	due bank
157	期中付款	interim payment
158	付款期	payable period
159	收款人	payee
160	付款人	payer
161	凭发票付款	payment on invoice
162	终止时的付款	payment on termination
163	支付条款	payment provisions
164	支付计划,支付进度表	payment schedule
165	折旧年限	period of depreciation
166	宽限期	period of grace
167	定期付款	periodic payment
168	特惠津贴	exgratia

第三节 对话口译

Dialogue 1：在工程验收会议上，业主和承包商代表的一次对话（音频：10.3.1）

Word Tips

acceptance time	验收时间
cost overruns	超支
maintenance services	维护服务
surface reinstating	恢复地表面

【场景：At the project acceptance meeting, the owner and the project manager talked about the project acceptance.

在工程验收会议上，业主(以下简称 A)和项目经理(以下简称 B)就工程验收事宜展开谈话。】

A：尊敬的项目经理，欢迎您参加我们的工程项目验收会议。请您确认我们的验收时间和地点是否符合您的要求。

B：Thank you very much. The time and location are suitable. Let's begin the acceptance process.

A：好的。我们先来看项目的进度。请您说明是否在预定的时间内完成了工程建设。

B：Yes, the construction work was carried out as planned and substantially completed. But there are minor items, like surface reinstating, etc.

A：我的意见是，验收分两个阶段：第一阶段，在你方技术人员的协助下，我们的专家组检查已完工的工程，看看是否令人满意。如果一切正常，我们进入第二阶段，也就是对合同中规定的设备和仪器进行性能试验。

B：I agree.

A：接下来，我们来谈一下工程项目的经济效益。请您介绍一下工程实施过程中是否发生了超支或节约情况，并解释原因。

B：During the implementation, we closely controlled the costs and there were no instances of cost overruns. In certain areas, we also took reasonable measures to achieve cost savings.

A:非常好。最后,请您介绍一下工程项目的可持续性和环境影响。

B:We fully considered sustainability and environmental protection factors in the project design and implementation. We utilized low-carbon, environmentally friendly technologies and materials to minimize the impact on the environment.

A:非常感谢您的介绍。根据我们的讨论,我可以确定这个工程项目符合我们的要求和期望。非常感谢您和您的团队的辛勤付出。

B:Thank you for your recognition. We also greatly appreciate your support for our work. If needed, we will be available to provide ongoing technical support and maintenance services.

Dialogue 2:关于支付方式的对话（音频:10.3.2）

Word Tips

60-days credit period	60 天信用期
L/C	信用证
freight documents	货运单据
pay in installments	分期付款
full payment	全额付款
hard currencies	硬货币
advance payment	预付款

【场景:Mr. Liu is the purchasing department manager, and Mr. Yang is the seller. The following is a conversation between them regarding payment methods and details.

刘先生是采购部经理(以下简称 L),杨先生是卖方(以下简称 Y)。以下是他们双方沟通付款方式及付款细节的对话。】

Y:我们希望贵方用美元支付。

L:We hope you can accept payment in other currencies except in US dollars. Would you agree to a 60-days credit period?

Y:抱歉,我们要求 45 日内付全款。如果贵方是用现金付款,我们给九折优惠。

L: But payment by L/C is our method of trade in such commodities.

Y: 在收到货款的十天内，我们就可以把货送出去。

L: We prefer payment after delivery, because these goods are very expensive.

Y: 我方要求贵方在收到货运单据后，立刻支付货款。

L: Because of the money problem, I hope that you can allow us to pay in installments with the first payment after delivery, and then we'll pay the rest once a month.

Y: 很抱歉，我们不支持分期付款，我们希望能收到全额付款，因为我们也需要付钱给厂家。

L: Could you make an exception in our case and accept D/P or D/A?

Y: 我方遗憾地告诉贵方，不能考虑贵方用承兑交单的方式来支付款项的请求。你们可以先用现金付定金，剩下的钱45个工作日内结清即可。

L: We can only accept 20% cash payment in local currency. The other 80% by L/C should reach us 15 to 30 days before the delivery.

第四节　篇章口译

Passage 1:对颁发完工证明的解释（音频:10.4.1）

Word Tips

completion certificate	完工证明
potential defects	潜在缺陷
comprehensive acceptance evaluation	全面的验收评估

【场景:The owner refuses the contractor's request to issue a certificate of completion and explains it accordingly.

业主方拒绝承包人提出颁发完工证明的要求，并做出相应解释。】

Dear representative from the contractor's side, regarding your request for issuing a completion certificate, we regret to inform you that we are unable to agree to this request. I will provide a detailed explanation of our reasons.

Firstly, according to the agreement in our contract, a completion certificate can only be issued after the formal acceptance of the project. However, we have not yet conducted the formal acceptance of the project, and as such, we are unable to provide a completion certificate at this time.

Secondly, a completion certificate is an important document and represents the formal completion and handover of the project. As the client, we need to ensure that the quality, safety, and economic benefits of the project meet our requirements and expectations. Only after conducting a comprehensive evaluation and acceptance of the project can we accurately determine whether it meets the contractual requirements and decide whether to issue a completion certificate. Otherwise, issuing a completion certificate prematurely poses certain risks and may result in the discovery of issues or potential defects after the acceptance, causing unnecessary losses and risks for the client.

Lastly, we aim to maintain the rigor and impartiality of the procedures and processes specified in the contract. By making the decision to issue a completion certificate based on a comprehensive acceptance evaluation, we can ensure the quality and safety of the project and protect the legitimate rights and interests of the client.

In summary, due to the project not having undergone formal acceptance yet, we are unable to fulfill your request for issuing a completion certificate. We hope you can understand our position and are willing to cooperate with us in the formal acceptance of the project. Once the project passes the acceptance and meets the contractual requirements, we will promptly issue the completion certificate and ensure a smooth handover of the project to you.

Passage 2:土木工程合同中的付款流程和须知（音频:10.4.2）

Word Tips

进度款支付	progress payment
完工款支付	completion payment
土方工程	excavation
进场	mobilization
施工方	construction party
资金流动	financial flow

补充事项	supplementary matters
发票要求	invoice requirements
可追溯性	traceability
正规发票	regular invoice
违约	default
罚款金额	penalty amount
执行方式	execution met
arbitration	仲裁
weigh	权衡,考虑
postpone	推迟,延期
confirm	保兑
outstanding	未偿付的

付款方式

1. 预付款

在土方工程合同中,双方可以约定在施工开始前支付一定比例的预付款。这有助于施工方及时进场,并预付部分劳动力和材料费用。预付款金额可以根据工程规模和双方协商确定,但一般不超过合同总额的30%。

2. 进度款支付

土方工程的付款方式中,根据工程进度付款是常见方式之一。当工程达到一定进度时,业主按照约定的比例支付相应的款项给施工方。这种方式有利于激励施工方积极推进工程进度,保证项目按时完成,同时也避免了施工方资金流动的问题。

3. 完工款支付

当土方工程完工时,业主需要支付剩余的完工款给施工方。完工款通常是扣除预付款和进度款之后的尾款,也是施工方最后获得的报酬。合同中应明确约定完工款的支付时间和方式,确保施工方得到相应的回报。

有关付款方式的补充事项

1. 发票要求

在土方工程合同中,双方应约定付款时的发票要求。明确要求施工方提供正规发票,以确保合法性和可追溯性。

2. 扣款原因和方式

当发生违约或者质量问题时，业主有权扣除相应的款项。因此，在合同中需要明确约定违约和扣款的条件、原因以及方式。

3. 罚款金额和执行方式

如果施工方未能按照合同约定的时间完工，业主有权要求施工方按照约定支付罚款。合同中应明确罚款金额和执行的方式。

第五节　口译技能——模糊语口译

语言的模糊性是相对于精确性而言的。二者皆为自然语言的本质属性。所谓模糊性是指客观事物差异在中介过渡时所呈现的"此亦彼性"，也指能够表达一定的内涵以及一些特定的事物，却又没有明确的外延，表达的只是模糊的概念，即属于某一范围的事物却又不太明确的概念。如"富有"，我们无法明确地界定有多少财富属于富贵、有多少财富属于贫寒，原因就在于其定义边界模糊。

自然语言存在模糊语言，主要有两个原因：一是自然界的客观属性包含模糊性，例如高与矮、胖与瘦、大与小等；二是人类的认知能力在一定的时期内是有限的，如在丰富奇妙的大自然中，生物形态各异，人类难以用现有的知识与语言去区分以及描述每一个生物的特性和名字。

口译是一种高强度、临场性的工作，因此要求译员在会场受到干扰的工作环境下迅速完成"听懂—理解—速记—信息重组—表述"这一过程——从说话人发言停止到译员开口的时间一般不超过 5 秒。因此无论遇上什么困难，口译员都要迅速将听到的信息传译出去。在大多数情况下，各种临场因素的制约，如设备扩音过大、信号干扰、设备产生嘶鸣声等，都会导致译员听不清、听不懂、突然卡壳或找不到对应词语等。此时，在时间极为紧迫、精神极度紧张的情况下，口译员可以适当使用一些模糊表达法来缓解自身压力，争取更多的时间来思考，为听清下一句做心理准备，以退为进，从而顺利完成口译工作。

一、利用模糊表达，争取思考时间

例如这句话："偌大个中国要解决的事有很多，如果说主要问题的话，首位还是发展经济"。现场口译为"In a country as vast as China. we have a lot of tasks to fulfill. As for the most important one. I think it is to facilitate the continuous growth of the economy. ",此句话的翻译很通顺流畅并且译员使用了"I think"这一主观判断的限定语。该模糊表达既使句意更加通畅,也为口译员争取了宝贵的时间,为下句话的翻译提供了良好的铺垫。所以模糊语的使用在口译当中是必不可少的一个技能,但该技能需要反复操练才能加以应用。

二、数字模糊表达,务必保证精准性

在口译实践中,由于汉英两种语言的差异性,很多情况下如外事谈判、经贸对话、贸易磋商等,都会涉及一连串数字的翻译,然而中西方在数字表达上有着非常大的差异。例如:汉语中的万、亿等在英文中就没有对应的单位,只能记为10 thousand、100 million 等。并且汉语是四位计数,英文是三位。这样的计数方法就给译员在现场翻译时带来了极大的困难,不仅听记困难,重述转换也困难。如在政府工作报告中连续听到一连串的数字,译员可能来不及记下所有的数字,例如975,323,468,530.04,与其漏译、误译,不如记下字头9753 亿,在翻译时加上模糊限制语,如大约、大概、多、以上等,务必保证精准性。数字口译的练习是一个持续而不间断的过程,需要译员花费大量的时间、精力。只要勤加操练,他日定有所成。

三、专有名词模糊化,避免引起反感

我们生活的大千世界是无奇不有的,原则上口译员必须熟记常用名称、地点、工具等。但口译员由于年轻、经验不足,很难熟知众多分支学科、科研领域,专有名词记不全、记不牢的事情时有发生。当这种情况发生时,如未能准确记下某名词,可以利用概括定义的方法对其进行描述,以便使专有名词模糊化。例如,当听到"涡轮增压发动机"时,其专有名词为 turbo supercharged engine,若一时没有记下,便可以将其描述为:"该空气压缩机利用发动机的废气为动力,从而增加发动机里面的空气。"这样的概括描述也不为过。这样处理是为了避免因误译而被认为不尊重别国人民和文化,以免引起听众反感。

由于受临场性、高强度的工作环境所制约,口译没有充足的时间来推敲琢磨。当工作时间过长、精力不济时,译员就可能会出现误译、漏译等现象,于是听众便很难接收到准确完整的信息。因此口译中模糊语言的使用,可以有效地减轻译员的临场压力,从而为随后的翻译做好铺垫,避免因漏译、误译产生的歧

义降低口译质量。尽管模糊语言的应用是十分有必要的,但译员却不能以模糊为追求目标和翻译准则。因此在口译过程中仍需始终坚持以准确、通顺、流畅为标准。

第六节　课后练习

(音频:10.6)

Ⅰ. Technical Terms Interpretation

1. 实质性竣工

2. 工程竣工验收

3. 质量监督员

4. 专门小组

5. 技术员工

6. 移交证书

7. 项目启动

8. 验收动态

9. 实施标准

10. 项目稽查

11. cash against delivery

12. cash on delivery (COD)

13. cash on shipment(COS)

14. cash with order (CWO)

15. delayed payment

16. out-of-date check

17. outstanding check

18. paid check

19. audited financial statement

20. balance sheet

II. Sentences Interpretation

1. 我们已经为项目的验收准备了一份计划。

2. 我们会尽快完成未尽的项目。

3. 我是验收方的正式授权代表,有问题请与我联系。

4. 只有交付物满足合同中的具体标准,业主才会同意接收。

5. 在大型项目中,期望零问题报告一般是不合理的。

6. 请注意,此笔款项是一次性特惠津贴。

7. 多退少补。

8. 如果进一步拖延付款的话,我方将被迫依照合同采取相应的行动。

9. 这是我方认为根据合同规定应支付给我方的估算额。

10. 请告知我方该银行的名称、账户名和账号。

III. Dialogue Interpretation

Word Tips

D/A payment	承兑交单
D/P	付款交单
irrevocable letter of credit	不可撤销信用证
L/C(a letter of credit)	信用证
deposit	押金

【场景:Mr. Smith is the buyer and Mr. Ma is the seller. Mr. Smith and Mr. Ma have already confirmed the goods, and there is currently a partial conversation about payment methods.

史密斯先生是买方,马先生是卖方。他们已经确定好商品,这是目前关于付款方式的部分对话。】

Mr. Smith:好吧,既然价格、质量和数量问题都已谈妥,现在来谈谈付款方式怎么样?

Mr. Ma: We only accept payment by irrevocable letter of credit payable against shipping documents.

Mr. Smith:我明白。你们能不能破例接受承兑交单或付款交单?

Mr. Ma: We can't agree to D/A payment terms. It is only when the trading parties know each other very well and there doesn't exist any risk of non-payment, will D/A payment terms be adopted. If goods are shipped and then rejected on arrival at the port of destination, we would be put to no end of trouble and financial losses.

Mr. Smith:那怎么可能呢？你们应该相信我嘛,对不对？老实说,信用证会增加我方进口货的成本。在银行开立信用证,我得付一笔押金。这样会占用我的资金,从而增加成本。

Mr. Ma: Consult your bank and see if they will reduce the required deposit to a minimum.

Mr. Smith:对不起,这种情况下,承兑交单或付款交单是最好的付款方式。

Ⅳ. Passage Interpretation

我方在 2023 年 4 月 8 日的函中指出,如果该函收到后 7 天之内未全额付清付款,我方将暂时停止履行我方的合同义务,但至今我方仍未收到任何(全额)付款。作为直接后果,现谨通知你方,我方暂停履行我方所有的合同义务。

最终付款证书本应于 2024 年 8 月 14 日签发。自此日期已逾两周,而我方仍未收到付款证书。

在不影响我方在此问题上的权利的前提下,如果在 2024 年 9 月 14 日前能收到最终付款证书,则我方将不会对这一违约行为采取进一步的行动。

随函附上我方根据 2024 年 11 月 10 日的会议上双方同意的新的计价方法完成的合同总价调整的计算书。

第十一章　工程项目索赔

"索赔"一词来源于英语"claim",其原意表示"有权要求",法律上叫"权利主张",后引申为"索赔"。索赔是一种正当的权利要求,它是业主、监理工程师和承包人之间一项正常的、大量发生而且普遍存在的合同管理业务,是一种以法律和合同为依据、合情合理的行为。

第一节　背景知识

一、索赔的概念及分类

(一)索赔的概念

索赔是指在工程承包合同履行中,合同当事人一方由于另一方未履行合同所规定的义务而致使本方遭受损失时,要求对方给予赔偿或补偿的权利。索赔的发生是双向的,只要合同中一方的责任和义务未按合同约定实现,或出现提供的条件与合同的约定不一致,就有可能出现索赔。

(二)索赔的分类

工程索赔按目的可分为两类:工期索赔和费用索赔。

工期索赔是指在工程施工中,由于非承包人的原因而导致施工进程延误,要求批准顺延合同工期的索赔。施工单位提出工期索赔的目的通常有两个:一是免去或推卸自己对已产生的工期延长的合同责任,使自己不支付或尽可能少支付工期延长的罚款;二是进行因工期延长而造成的费用损失的索赔。如果工期延长不是由施工单位造成的,而建设单位已认可施工单位提出的工期索赔,则施工单位还可以提出因采取加速措施而增加的费用索赔。费用索赔以补偿实际损失为原则,其目的是要求经济补偿。当现实的条件与合同的约定不一致,导致承包商增加开支,可以要求对超出计划成本的附加开支给予补偿,以挽回不应由承包商承担的经济损失。

二、索赔的原因

(一)合同签订不严谨引起的索赔

合同是由合同协议书、招标文件、投标书、合同专用条款、合同通用条款、图纸、工程量清单及合同履行过程中的补充协议等一系列的文件所组成的,经业主和承包商双方依法签订生效,具有法律约束力,任何一方不得擅自变更、解除或不履行合同赋予的权利和义务。但工程项目建设的复杂性和自然环境、气候、周期长等因素的限制,加上合同用词不严谨、文件之间相互矛盾等,都有可能使双方在签订施工合同时没有充分考虑和明确各种因素对工程建设的影响,从而引起施工索赔。

(二)工程设计方面引起的索赔

施工图纸存在缺陷或错误,施工图与现场实际施工在地质、环境等方面存在差异,设计的图纸与规范要求不符,施工说明表达不严谨,设备、材料的名称、规格、型号表示不清楚,工程量错误等都可能造成返工,从而不可避免地产生工期、人工、材料等方面的索赔要求。

(三)管理模式引起的索赔

当前的建筑市场,工程项目建设承发包有总包、分包、指定分包、劳务承包、设备和材料供应承包等一系列的承包方式,这使工程项目建设承发包变得复杂,管理模式难度增大。任何一个承包合同不能顺利履行或管理不善,都会影响工程项目建设的工期、质量和数量,继而引发工期、质量、数量和经济等方面的索赔。如设备、材料供应商不按工程项目的施工进度和设计要求按时保质提供设备、材料,工程也就不能按业主的要求和设计的质量以及有关规范的要求进行施工,从而影响工程项目建设的进度和质量,最终导致业主、总包方、分包方、设备材料供应商相互索赔。

(四)不可抗力引起的索赔

在施工过程中,施工现场条件的变化对工期和造价影响很大,如地震、台风、战争、放射性污染、核危害以及在施工中出现地质断层、天然溶洞、沉陷、地下文物或构筑物等地下不明障碍物等。这些都是人力不可抗拒的客观因素,常常导致工程变更,引起施工索赔。如在挖方工程中发现地下构筑物和文物等,图纸上并未说明,确属施工单位难以合理预见的人为障碍,如处理则必然导致工程费用增加,施工单位即可提出索赔。

第二节　相关词语与表达

序号	中文	英文
1	承认	acknowledge
2	投诉	complaint
3	赔偿	compensation
4	损害	damage
5	延迟	delay
6	不满意的	dissatisfied
7	错误	error
8	交换	exchange
9	有缺陷的	faulty
10	保证	guarantee
11	不便	inconvenience
12	调查	investigation
13	责任	liability
14	实际损失	actual loss
15	合同补遗	addendum to the contract
16	协助运送	aid the transit
17	没有达到标准	be not up to the standard
18	感到失望	be upset
19	以仲裁方式	by way of arbitration
20	召开一个会议	call a meeting
21	向承运人索赔	claim against carrier
22	损伤索赔	claim against damages
23	向保险商索赔	claim against the underwriters
24	索赔计算书	claim calculations
25	索赔不予考虑	claim cannot be entertained
26	要求赔偿	claim for indemnification
27	要求偿还	claim for reimbursement
28	退休金请求权	claim for retire pay

续表

序号	中文	英文
29	侵权索赔	claim in tort
30	多式联运索赔程序	claim procedures in multi-modal transport of goods
31	索赔报告	claim report
32	索赔清单	claim statement
33	检验代理人	claim surveying agent
34	索赔人	claimant
35	被索赔人	claimee
36	索赔货物	claimed goods
37	索赔代理人	claims agent
38	估损人	claims assessor
39	申诉审理委员会	claims commission
40	理赔部	claims department
41	索赔证件	claims documents
42	赔偿估算，赔偿预算	claims estimated
43	非限制性债权	claims excepted from limitation
44	理赔费用	claims expenses
45	非限制性债权	claims not subject to limitation
46	索赔通知	claims notice
47	赔偿付讫	claims paid
48	国外可付索赔	claims payable abroad
49	铁路运输索赔	claims procedure under railway contracts
50	海洋污染损害索赔报告书	claims report on marine pollution damage
51	船舶(海事)优先权担保的债权	claims secured by maritime lien
52	理赔代理人	claims settling agent
53	理赔代理费	claims settling fee
54	索赔清单	claims statement
55	同期记录	contemporary records
56	来往函件	correspondence
57	费用补偿	cost reimbursement
58	弥补我们的全部损失	cover all our losses

续表

序号	中文	英文
59	不同的现场条件	differing site conditions
60	进行直接的谈判	engage in a direct negotiation
61	不可抗力事件	events of force majeure
62	公平合理的	fair and equitable
63	经历许多困难	go through many difficulties
64	理赔	honor the claim
65	国际建筑工程承包	international construction contracting
66	危害,危及	jeopardize
67	漫长的商谈	lengthy discussion
68	定期维护	maintain regularly
69	向……索赔	make claim against
70	条件是……;以……为条件	on condition that
71	解决问题	resolve the issue
72	索赔的金额	the amount claimed
73	索赔局	claiming administration
74	要求补偿	claim for compensation
75	由于损坏而索赔	claim for damage
76	要求索赔	claim for indemnity
77	贸易纠纷(引起的)索赔	claim for trade dispute
78	由于短重而索赔	claim for short weight
79	理赔代理费	claims settling fee
80	解决	in settlement of
81	无能力	inability
82	由于质量低劣而索赔	claim for inferior quality
83	放弃索赔	to waive a claim
84	撤销索赔	to withdraw a claim
85	仲裁庭	arbitral tribunal
86	罚金	penalty
87	保险索赔	insurance claim
88	延期装运索赔	claim on delayed shipment

续表

序号	中文	英文
89	提出索赔	to make a (one's) claim
90	提出索赔	to register a (one's) claim
91	提出索赔	to file a (one's) claim
92	提出索赔	to lodge a (one's) claim
93	提出索赔	to raise a (one's) claim
94	提出索赔	to put in a (one's) claim
95	提出索赔	to bring up a (one's) claim
96	拒赔	claims rejected
97	就某事提出索赔	to make a claim for (on) sth.
98	向某方提出索赔	to make a claim with (against) sb.
99	不利的障碍或条件	adverse physical obstruction or conditions
100	友好解决	amicable settlement
101	索赔额	amount of claim
102	仲裁协议	arbitration agreement
103	仲裁解决	arbitration award
104	索赔评定	assessment of claims
105	一揽子索赔	blanket claim
106	索赔额外费用	claim for additional payment
107	索赔权	claim right
108	利益的冲突	conflict of interest
109	承包商的立场	contractor's position
110	承包商的请求	contractor's request
111	变更意向通知	contemplated variation notice
112	有争议的双方	contesting parties
113	意外事件	contingency
114	合同规定	contract provisions
115	合同工作范围	contract scope of works
116	承包商的意见	contractor's opinion
117	合同规定的索赔	contractual claims
118	费用的减少	decrease in cost

续表

序号	中文	英文
119	减少合同工程量	decrease the quantity of the contract
120	违约方	default party
121	缺陷责任	defect liability
122	设计变更	design change
123	详细索赔	detailed claim
124	变更费率或价格	vary a rate or price
125	争议评审团	dispute review board
126	听证会	hearing
127	有效合同价	effective contract price
128	工程师的决定	engineer's decision
129	工程师的意见	engineer's opinion
130	工程师的立场	engineer's position
131	工程师的建议	engineer's proposal
132	业主的立场	employer's position
133	当面谈判	face-to-face negotiation
134	未能遵守	failure to comply
135	最终索赔	final claim
136	费用的增加	increase in cost
137	增加合同工程量	increase the quantity of the contract
138	增加的费用	increased cost
139	变更指令	instruction for variation
140	向承包商下达指令	instruction to contractor
141	索赔意向	intention to claim
142	中间索赔	interim claim
143	国际商会	international chamber of commerce
144	发布指令	issue an instruction
145	会议纪要	minutes of meeting
146	索赔通知	notice of claims
147	违约通知	notice of default
148	致承包商的通知	notice to contractor

续表

序号	中文	英文
149	省略工作	omit work
150	口头指令	oral instruction
151	管理费费率	overhead rate
152	索赔的支付	payment of claims
153	仲裁地点	place of arbitration
154	潜在争议	potential dispute
155	工程师确定费率的权利	right of engineer to fix rates
156	索赔程序	procedure for claims
157	适合变更工作的费率或价格	rate or price applicable to the varied work
158	变更原因	reason for variation
159	费用记录	records of costs
160	相关条款	relevant clause
161	重新估价	revaluation
162	仲裁规则	rules for arbitration
163	争议的处理	settlement of dispute
164	值班日记	site diary/journal
165	现场记录	site record
166	后继法规	subsequent legislation
167	索赔证明	substantiation of claim
168	工时记录	time keeping
169	变更估价	valuation of variation
170	工程量变更	variation of quantity
171	变更令	variation order

第三节 对话口译

Dialogue 1: Negotiation on the Claims Caused by Flooding（音频：11.3.1）

Word Tips

contemporary record	同期记录
reimbursement	补偿
claim calculations	索赔计算
final offer	最终报价
equitable	公正的

【场景：Due to flooding, construction progress was hindered and the project suffered significant losses. Therefore, the Chinese contractor claims compensation from the external owner, and both parties send their representatives to negotiate.

由于发生洪水,施工进度受阻,且项目建设遭受极大损失。为此,中方承包人(以下简称 C)向外方业主(以下简称 E)索赔,双方派出他们的代表进行谈判。】

E:Despite the great difficulties, the project is now progressing smoothly. On behalf of my company, I'd like to extend my sincere gratitude to your great effort. I hope we can work together to find a solution that is acceptable to both sides.

C:说真的,我们对你方的反应感到失望。6 月 12 日和 24 日的两次洪水使我们遭受了很大损失,这一点,您可以从我们的现场同期记录和索赔计算中看出。如果我们得不到理赔,这将很可能不利于项目的顺利进行。

E:I had studied all the correspondences regarding your request for cost reimbursement and schedule extension before I came here. It is a fact that you have suffered a loss because of the high water levels; it is also a fact that the loss is partly due to your inappropriate site activities. I regret for what happened. However, you're asking too much.

C:我确信,我们关于费用和工期补偿的要求是合理的。不管怎么说,我想先听听您的意见。

E:Let me talk straight. We allow you two-month schedule extension and a total

— 213 —

sum of US $100,000 as compensation for all of your losses. This proposal represents our final offer. I believe it is fair and equitable.

C:感谢您提出的这一方案。2 个月的延期我们能够接受,但你方的费用补偿大大低于我们的索赔额,而这一金额是基于我们的实际损失得出的。

E:I have to say we have done our best. To settle this claim, both sides have to make concessions.

C:我承认你方迈出了一步,但步子太小。我们始终希望能与你方保持良好的工作关系;我们始终希望能够向你方交付成功的项目,但我们也希望得到公平合理的报偿。如果必要,我们可以通过仲裁来解决这项索赔。

E:Well, let me suggest at this point that we increase the compensation to US $150,000. This proposal really does represent our final offer. If the offer is not acceptable to you, we have no choice but to refer the matter to arbitration.

C:本着相互理解和利于我们将来合作的精神,我们接受您的这一方案,尽管这一金额远不能弥补我们的损失。

E:I support your call for mutual understanding and cooperation. I'll draft today's agreement as an addendum to the contract and forward it to you for your review and signature.

第四节　篇章口译

Passage 1: A Speech on the Insights and Experience Summary of Construction Claims Management（音频:11.4.1）

Word Tips

工程延期	project delays
会议记录	meeting minutes
设计变更通知	design change notices
工程日志	project logs

【场景:The following is a speech delivered by a project manager on the insights and experience summary of daily management of construction claims based on his own daily work.

以下是一位项目经理结合自身日常工作就施工索赔的日常管理工作的见解和经验总结发表的讲话。】

尊敬的各位领导、亲爱的同事们：

大家好！在项目管理的过程中，我们常常会面临各种挑战和困难，其中之一就是索赔管理。索赔是项目过程中不可避免的一部分，我们需要以积极的态度和专业的方法来处理。我想强调的是，索赔并非一定是负面的体验，而是可以视为项目管理中的一次学习和提高的机会。

首先，我们需要了解索赔的本质。索赔通常是由项目中的一些不可预测的风险和变更引起的，这可能包括合同条款不明确、设计变更、工程延期等因素。索赔并不仅仅是一种纠纷或争端，它是建立在合同基础上的法律要求，是对在工程项目中出现的预期之外的情况所导致的经济或时间损失的一种申请。理解这一点可以帮助我们更客观地看待索赔，不仅是从法律的角度出发，更要考虑到背后的实质问题。作为项目经理，我会在项目计划和合同中尽量考虑到这些因素，以降低索赔的风险。但是，即便我们尽最大努力，有时候仍然无法避免索赔的发生。

对于索赔的处理，我们应该采取一种积极主动的态度。首先我们需要认真审查合同条款，确保我们有权拥有并行使相关的索赔权利。与此同时，我们也要确保我们的索赔是合理、充分的，有充分的文档和证据支持。这些文档和证据包括但不限于合同文件、会议记录、设计变更通知、工程日志等。

在提出索赔之前，我们需要与相关方进行充分的沟通和协商。沟通是解决问题的关键，通过与业主、设计方、承包商等相关方进行有效的沟通，我们可以更好地理解彼此的需求和立场，找到解决问题的最佳途径。在沟通的过程中，我们需要保持冷静、理智，避免情绪化的表达，以确保沟通顺畅和有效。

另外，我们也需要注意索赔的时机。及时提出索赔可以避免问题进一步恶化，并有助于更早地解决问题。在提出索赔之前，我们需要对索赔的影响和可能的解决方案进行充分的评估，确保我们的索赔是合理的，有助于项目的顺利进行。

最后，索赔处理过程中，我们需要与团队保持良好的沟通和协作。团队的支持和合作是解决问题的关键,我们需要鼓励团队成员提出建议和意见,共同努力找到最佳的解决方案。通过团队的努力,我们可以更好地应对各种挑战,确保项目能够按计划顺利推进。

总的来说,索赔管理是项目管理中的一个复杂而重要的方面。通过积极主动的态度、充分的准备和有效的沟通,我们可以更好地处理索赔,并从中吸取经验教训,提高项目管理水平。感谢大家的合作和支持,让我们共同努力,确保项目成功完成。

谢谢!

Passage 2：Response to Claims（音频：11.4.2）

Word Tips

| embedded pipe | 预埋管 |
| foundation slab | 筏基 |

【场景：The following is the response of the owner to the claim made by the contractor.

下文是业主针对承包人所提索赔做出的回应。】

Your claim is about the embedded pipes in foundation slab of pump house, and the materials request in the construction drawings are different from the technical specification in the Contract Documents. You have submitted several technical clarification requests and been replied till 35 days later. You hold that the final decision has been made too late and it has made some loss in cost and construction period, and that a compensation for time extension and cost increasing is necessary.

But in our opinion this claim is not effective. You will understand it after we review the claim in detail in accordance with the procedure for claims. Firstly, you have submitted the claim within the time regulated in the Condition of Contract. Secondly, you also submitted the bill of the cost of the claim within 30 days after the first instruction. But after we requested the further analysis for the claim event according to the procedure mentioned in the Conditions of Contract, you failed to submit necessary information within 14 days. That means you are not able to provide enough evidence for the claim.

You have put forward clear requests in your claim: point 1, compared with the drawing schedules, the issuing of construction drawings and the technical clarification have delayed for 42 days. For catching up the original schedule you have mobilized more material, machines and man-power which cost USD 25,397; point 2, the price of the new material in the new drawings is USD 15,983 higher than that in the Contract document. But we cannot agree that your description can be used as the cost analysis. We can only accept what comes from the proper procedure. In this case we have to say that this claim is not workable and therefore it is not effective.

第五节 口译技能——工程口译的应急处理

工程现场口译不同于笔译,口译人员必须及时、独立地一次性翻译。工程现场口译人员受到时间的限制,必须以最快的速度传递双方交流的信息,译员没有足够的时间查字典或其他资料,也不可能在现场仔细推敲或向有经验的同行求教。在工程现场,由于受到双语甚至多语种的专业知识和翻译技能的限制,加上现场心理压力大,初涉工程现场的口译者往往会出现听不清、听不懂、译不出或译错的情况。这就要求译员除具备扎实的语言功力和专业知识外,还必须具备一定的应急处理技巧,在紧急情况下采取补救措施。

一、源语难以理解

口译时,最怕遇到听不懂的词。克服这一障碍的唯一办法是培养自己的猜测能力和预测能力。当然,事先充分了解必要的背景材料和知识是十分重要的。这样做,译员就能心中有数,知道讲话人要谈什么方面的内容。译员在这个基础上进行翻译,即使遇到个别不会的词,根据上下文,或根据对整个讲话的体会,也可以将全句内容猜测出来。如果遇到不懂的句子,可跳过去,不要因为一句话没有听懂而影响整个上下文的翻译。

在工程现场遇到重要词语听不清或听不懂时,切忌慌乱或凭感觉任意猜译,亦不可删减遗漏,否则会导致现场管理失误。一般情况下,对于不懂之处,译员应请求讲话者进一步解释、澄清原意,也可在准确理解后利用手势、面部表情、眼神等方法转达讲话者所传递的信息。

具体说来,在处理工程口译实践中听不懂的词语时,可采用以下一些方法:

1. 求助法

工程现场口译的语言特点是专业词语和省略句多,口音、方言以及语言变体多。由于地理环境、文化背景、个人素养的不同,不同的人所用的语音、语调、词语和语言习惯等也各有差异。在现场负责施工、安装、调试的工程技术人员可能来自不同的国家和地区,他们在讲英语的时候往往会受到本国语言发音的影响。这些因素都给译员的翻译增加了难度。例如:"a"在英式、美式英语中发[ei]的音,而在澳大利亚却更接近[ai]的音,如"day[dei](日子)"说成[dai](音同"die")。在翻译时听不懂或听不清时,正确的做法是请说话者澄清原意,然后再转达说话者所要表达的信息。常用表达方式有"Pardon!""I beg your pardon!""Sorry, what do you call(say)... in English?""What do you mean by...?"。

此外,也可采用间接方式求助,如用眼神、手势、面部表情、重音、语调等。但求助法不可用得过多,在工程现场,时间不允许你总让对方重复后再译。若总是说"Sorry"或者"Pardon"之类的话,一个信息要花很长时间才能传递过去,会令交流的双方处于等待之中,这样不仅会影响工作进展,甚至还会令人对译员的素质产生怀疑。

2. 迂回法

迂回法就是在不能或不会准确表达时巧妙地绕开障碍,进行解释性翻译。这样做虽烦琐,但能达到顺利翻译之目的。在现场口译时,常常碰到许多专有名词,如炼胶工艺中的"发黑(carbon black)""塑炼(plastication)""静电除尘器(electrostatic precipitator)"等晦涩难记的专业词语,给现场口译增加了难度。比如,在检测供电设施时中方人员问外籍人员:"你是否检测过所有的瓦斯继电器?"译员只熟悉"继电器(relay)"这个词,但"瓦斯"用英语怎样说一时想不起来。双方都在等待译员的翻译,在这种紧急情况下可采用解释性翻译法:"Have you inspected and tested all the gas relays? I'm sorry I'm not sure whether the term 'gas relay' is correct. By 'gas' I mean a kind of combustible gas. (你是否检测过所有的瓦斯继电器。抱歉对'gas relay'这个术语我不敢肯定。我说的'gas'是指一种易燃气体。)"如果外籍人员听后回答"yes",这就表明解释性翻译起了作用。实际上,"瓦斯继电器"的英文就是"gas relay"。称职的译员不是词句转换的"对号者",而是沟通思想的"搭桥人"。因此,他可以在无损原意的前提下进行词句的加工。在翻译一些中国特有的词或用语时,如与中国国情、政治、经济、社会、文化、名胜古迹等方面相关的词语,也可以进行迂回性的解释性翻译法。

3. 模仿法

为了使专业术语译得准确,使专业内容表达得正确,最好采用英汉词典上的有关术语及表达式。我国引进的设备可能来自非英语国家,因而一些技术术语或设备名称存在不规范现象。此时,译员不可将词典作为学习、寻查术语的唯一工具,或按照自己自以为正确的方式对某些专业性很强的术语进行解释翻译。这样做往往使翻译生硬、外行,甚至很可能会出现错误,达不到双方友好交流和协同工作的目的。译员要留心交流双方在现场工作中常用的专业术语和表达方式,口译时可使用模仿法,使自己尽快进入交流的环境。比如,在橡胶密炼机安装、调试工程现场,外方人员将"橡胶密炼机(internal mixer)"称为 mixer,将"胶片冷却系统(rubber cooling system)"称为"batch-off"等。译员则应在翻译中使用这些词语,使自己尽快适应现场环境。

4. 使用填补词法

口译人员在现场口译时,可能遇到一时找不到适当的词语来表达说话者意图的情况,为了使交流不致中断,可以利用填补词法来增加思考的时间。比如用"er""em""well"或"you see""you know""I mean"等。口译时,说话者不能准确表达或不需要表达十分确切时,也常用"about""around""kind of""sort of""somewhat"等词语,如"There are about 1500 workers. Oh, no. It's around 2 000 workers, that's right, in the plant."。

5. 借助实物法

与笔译相比,现场口译直观性较强,利用实物进行口译是最便利、最有效和直观的方式。例如,一位外籍程序调试人员在安排次日工作时说"I need a tape tomorrow."。这时,译员可能觉得奇怪:他要磁带干吗?但是看到他的计算机和针式打印机后就明白了,于是译为:"请明天给我带盒打印机色带。"在工程现场,译员还可使用代替词,如代词、指示代词、定冠词等。因为双方都在场,替代词既直观又不会引起误解。如"That meter should replaced.(这个仪表应换掉)"。

以上介绍的几个策略仅供偶尔应急补救使用。要解决现场口译中的障碍仍然要靠口译人员平时收集有关材料、了解相关科技领域的基本知识、加强中外文基础知识训练,并在长期的实践中不断总结经验,提高口译技能,增强随机应变的能力。

二、口译中的人称问题

由于说话人往往以某公司或组织机构的名义讲话,这些陌生的公司或组织名称有时会给译员造成困扰。在这种情况下,译员可用"我们"或"你们"等代词来代替。比如"With our continued investment in technology, people and facilities, the future of Holset has never looked so bright. "可以译为"我们在技术、人员和设备上不断投资,我们的未来异常光明"。

三、现场纠正口译错误

在口译现场,如何大大方方地纠正错误也是一门学问。译员发现自己译错后,如果是小错,只要不影响大局,则不必纠正,可在接下来的译语中自然地换用正确的词语;如果发现自己翻译的内容出现了大的错误,必须马上纠正,可以明确地说"刚才这点翻译错了,应该译为……",千万不能顾及自己的面子而造成更大的损失。

第六节　课后练习

(音频:11.6)

Ⅰ. **Technical Terms Interpretation**

1. 索赔

2. 处罚条款

3. 调解

4. 样品,样本

5. 差异和索赔条款

6. 不良包装

7. 延期交货

8. 理赔

9. 索赔函

10. 撤回索赔

11. freight charge

12. goods in transit

13. a survey report

14. complaint

15. the bill of lading

16. implement duty

17. inferiority goods

18. goods in good condition

19. for one party's account

20. workmanship

Ⅱ. Sentences Interpretation

1. 通常情况是进口商没有收到他所期望的货物,货的质量或数量不符合合同标准,出口商不能按时收取货款,才会导致申诉与索赔。

2. 在合同履行中,如一方违约给另一方带来经济损失时,受损方要求违约方根据合同规定给予赔偿或补偿。

3. 处理损失方的索赔被称为理赔。

4. 有时损失不大时,受损方不一定会为了补偿而提出索赔。

5. 万一交付的货物不符合合同规定的质量、数量、包装等,买方应在重新检验的时限内提出索赔,索赔要出具一份由卖方认可的检验员签署的调查报告。

6. 换言之,尽管未能履行合同方支付了违约金,但其仍须履行自己的合同义务。

7. 当进出口双方卷入纠纷时,建议解决纠纷的方法是仲裁比诉讼好,调解比仲裁好。

8. 今早收到上述订单的货物。

9. 尽管大部分货物正确完好,遗憾的是 7 号货箱破损。

10. 据此,货物发生问题极有可能是在运输途中,故我方不承担任何责任。

Ⅲ. Dialogue Interpretation

【场景：Mr. Zhang, the project manager, is managing an engineering project that requires claims. He is seeking help from a professional project management consultant to inquire about some engineering project claims.

项目经理张先生所管理的工程项目出现了需要索赔的情况,他正在向专业的项目管理顾问寻求帮助,询问一些工程项目索赔的事宜。】

A：你好,我是项目经理张先生。有关我们项目的一些问题,我想了解一下关于工程索赔的情况。

B：Hello, Mr. Zhang. Do you have any specific questions or situations that I need to explain?

A：是的,最近我们注意到一些工程延误和额外的费用支出,我想了解一下是否可以进行工程索赔。

B：Oh, this is indeed an important issue. Can you specifically tell me which aspects have caused project delays and additional costs?

A：主要是在施工过程中发现了一些设计图纸上的不一致,导致需要调整工程和重新施工。此外,一些材料的供应也出现了问题,使得工程进度受到了影响。

B：I understand. In this case, we can consider filing an engineering claim. Firstly, we need to collect relevant evidence, including inconsistencies in the design drawings, specific reasons for project delays, and details of additional costs.

A：好的,我们已经开始整理相关的文件和数据。不过,在提出索赔之前,我想了解一下索赔的流程和可能的结果。

B：The process of claiming compensation usually includes submitting a claim application, approval and evaluation by relevant parties, and finally payment of the claim. During the application process, we need to provide a detailed explanation of the reasons for the claim and provide sufficient evidence to support our claim. The approval and evaluation process may take some time, but once approved, we can expect to receive corresponding compensation.

A：明白了,谢谢你的解释。我们会继续准备相关材料,并尽快提出索赔申请。

B：You're welcome. If you need any help during the preparation process, feel free to let me know. I hope everything is resolved smoothly.

Ⅳ. Passage Interpretation

For a moderate contract of general merchandise, a discrepancy and claim clause is often included. In case the goods delivered are inconsistent with the contract stipulations for quality, quantity, packaging, etc., the buyer should make a claim against the seller within the time limit of re-inspection and claim with the support of a survey report issued by a surveyor accepted by the seller. For a contract covering a huge shipment of goods or high value equipment such as a complete plant, a penalty clause might also be included and quoted when one party fails to implement the contract such as non-delivery, delayed delivery, etc. A penalty is often a certain percentage of the total contract value, but paying the penalty does not mean that the contract can be avoided. In other words, the party that has failed to implement the contract must carry out his contract obligations in spite of his payment of the penalty.

第十二章　工程技术人员后勤保障

后勤管理部门作为项目非生产性物资、职工福利保障、劳务服务等资源配置的协调部门，其首要任务在于解决项目职工的实际需求，所有后勤工作也都应该围绕这一需求开展。后勤管理涉及范围广，与项目整体和职工个体联系密切，对项目正常运行起到保障作用。

第一节　背景知识

后勤服务保障工作是施工企业日常管理的基础。随着施工企业规模的不断扩大，企业员工队伍越来越庞大。做好后勤服务保障体系的构建，能有效实现施工企业的稳健发展。

一、新时代建筑施工企业构建后勤服务保障体系的重要性

后勤服务保障工作是工程运转发展的基础，是保证工作质量、按期实现优良工程的一个重要方面。随着新时代社会经济的全面发展，人民对工程质量的要求日益提升，做好工程后勤服务保障工作具有重要的指导和现实意义。

首先，构建完善的后勤服务保障体系是施工企业深化转型升级、实现高质量发展的重要举措。后勤服务保障工作水平直接关系到施工技术人员的切身利益。例如，如果企业没有科学的后勤服务保障体系，那么施工人员就没有舒适的生活工作环境，甚至基本的福利待遇也得不到满足，久而久之就会出现积极性不高或懒散的现象，造成企业员工的工作活力不够，创新动力不足。

其次，加强后勤服务保障工作是增强施工企业部门相互沟通、树立良好企业形象的重要举措。随着施工企业规模的不断扩大，企业部门之间的沟通越来越少，而加强后勤保障服务工作可以为部门之间的沟通提供载体，加强人文关怀，从而形成相互合作、团结互助的企业文化氛围。例如，后勤保障部门通过举办各种文化活动可以有效地为不同部门的员工提供相互沟通的平台，活跃了氛

围,有效地促进了企业文化建设。

最后,加强后勤服务保障工作有助于降低企业成本支出,增强企业的经济效益。后勤工作在整个建筑工程企业中占据重要的地位,高效的后勤服务保障工作可以压缩企业的成本支出,使企业实现集约化发展。例如,通过规范就餐员工的就餐行为,消除浪费现象,可以减少食堂的成本支出等。

二、新时代建筑施工企业后勤服务保障体系建设所存在的问题

目前建筑施工企业后勤服务保障体系存在以下问题:一是后勤服务保障员工服务意识不强,专业技术水平不高,老龄化问题突出。人们对后勤服务保障部门的认识停留在传统的只提供后勤服务保障工作的层面上。例如,人们认为后勤部门属于"养老"机构,因此从事后勤服务保障工作的人员年龄偏大,学历水平不高。尤其是后勤服务保障人员大数据技术应用能力不足,导致后勤服务保障工作与职工需求存在一定的差异性,不能及时满足职工的需求。二是后勤服务保障制度不完善,市场化程度较低。完善的后勤服务保障制度是推动后勤工作的基础,但是目前施工企业的后期服务保障制度建设存在较大的不足,尤其是在各项制度执行上存在人情化的问题。例如,在食堂管理中对于铺张浪费的现象并没有严格按照相关的制度进行处理,导致各项制度形同虚设。三是缺乏与其他部门的沟通,协作意识不强。后勤服务保障人员缺乏主动与科技人员沟通交流的意识,主动沟通能力较弱。例如,接到报修等问题时,不能及时反馈具体情况或尽快落实维修工作。四是后勤服务保障工作宣传力度不够,职工对后勤工作认同感不强。

三、建立健全施工企业后勤服务保障体系的对策

(一)转变服务理念,增强专业技能

后勤工作关系到企业每位职工的切身利益,因此基于新时代发展的要求,施工企业后勤管理人员必须转变观念意识,改变传统的工作思维。例如,针对新时代的新要求,构建高效节能的施工模式是企业可持续发展的重要举措,而后勤服务保障部门是推动企业节能增效的关键。因此作为后勤管理部门工作人员必须从工作观念上转变,树立全心全意为职工服务的精神,将后勤保障工作作为践行企业价值观和社会主义核心价值观的具体体现。同时,后勤管理人员要加强学习,掌握最新的后勤管理技术。随着新时代的发展,大数据技术在后勤工作中的作用凸显,例如大数据技术在食堂管理中的应用增强了食堂管理的效益。因此,作为后勤管理人员必须加强学习,掌握应用大数据技术的能力。

(二)完善后勤服务保障机制,运用市场化管理模式

制度是保证各项工作顺利推进的关键,针对施工企业所存在的后勤服务保障机制不健全的问题,一方面要结合企业实际情况出台严格的后勤服务保障制度,加大对后勤服务保障制度的执行力度。例如,针对后勤服务保障制度执行不到位的问题,施工企业要针对突出问题细化管理制度,加强对后勤服务保障制度执行过程的监督和评价。另一方面,施工企业要进一步深化后勤服务保障改革工作,积极采取市场化运行模式,提高后勤服务保障工作的质量。施工企业可以通过引入市场化的运作模式,通过市场化激励手段激发工作人员的积极性。

(三)建立畅通的沟通渠道,加强与外界联系

沟通是提高后勤服务保障工作的重要手段,通过沟通不仅可以及时了解广大职工对后勤服务部门的意见,而且可以第一时间了解职工的需求,从而提高后勤服务保障工作的质量。结合实践调查做好沟通工作必须借助完善的信息沟通渠道,因此企业一方面要依托大数据技术、微信公众号等平台构建上下互动的沟通渠道。例如,针对微信的广泛应用,后勤服务保障部门可以构建后勤微信公众号,通过微信及时将职工的需求进行汇总与处理,同时通过微信公众号将后勤部门的工作成绩进行宣传,让职工了解后勤工作。另一方面,后勤部门要加强与外界的联系,通过与优秀后勤部门合作交流,及时将其先进的后勤管理经验引进来,以此提高企业的后勤服务质量和水平。

(四)强化内部管控,提升服务保障效能

一是在制度建设方面与优秀企业对标,使后勤管理制度流程规范化,重点修订完善合同风险控制、人力资源和绩效管理控制流程与管理制度,同时完善以全面预算管理为主线的管理体系,实现业务预算、资金预算、费用预算及投资预算与人、财、物等资源的匹配管理;二是严格执行各类支出规定,对建设工程、维修、食堂等后勤管理服务加强成本控制,增强职工的节约意识,加大管理力度,坚持高标准、低成本服务;三是贯彻落实廉政建设各项举措,后勤服务人员数量众多,综合素质参差不齐,要以高标准、严要求管理办事人员,对他们加强警示教育,做到警钟长鸣。

第二节 相关词语与表达

序号	中文	英文
1	人才	personnel
2	人事变动	personnel changes
3	协议	protocol
4	个人所得税	personal income tax
5	工伤	industrial injury
6	工作调动	job transfer
7	工资	wages
8	广播	broadcast
9	专家证	expert certificate
10	中秋节	Mid-Autumn Festival
11	介绍	introduce
12	公积金	accumulation fund
13	文件	file
14	文员	clerk
15	日历	calendar
16	出勤	attendance
17	失业补贴	unemployment subsidies
18	生育	bearing
19	申请	application
20	任命书	appointment letter
21	合计	total
22	合同	contract
23	安全	security
24	延期	delay
25	考勤	attendance
26	体检	physical examination
27	劳动	labor
28	医疗保险	medical insurance

续表

序号	中文	英文
29	岗位证	job certificate
30	护照	passport
31	报告	report
32	系统	system
33	补贴	subsidy
34	证件	papers
35	评价	evaluate
36	事故	accident
37	国庆节	National Day
38	姓名	name
39	实习生	intern
40	居留证	residence permit
41	录用	recruitment
42	招聘	recruit
43	规定	provide
44	试用期	probationary period
45	保险	insurance
46	养老金	pension
47	奖金	bonus
48	总经理	general manager
49	总金额	total amount
50	政策	policy
51	标准工资	standard salary
52	看板	bulletin board
53	研修	training
54	研修室	training room
55	退勤	retirement
56	除名	delisting
57	面试	interview
58	消防	fire protection

续表

序号	中文	英文
59	班长	monitor
60	缺勤	absenteeism
61	请假	leave
62	资格补贴	qualification subsidy
63	部长	minister
64	预算	budget
65	副总经理	deputy general manager
66	商谈	negotiations
67	培训	train
68	基本工资	basic salary
69	教材	teaching material
70	职务工资	position salary
71	寒暄	greetings
72	就业证	employment certificate
73	提案	proposal
74	联络	liaison
75	董事长	chairman
76	福利	welfare
77	签证	visa
78	履历表	resume
79	出厂合格证	factory certificate of conformity
80	材质证明单	material certification form
81	安全责任	safety responsibility
82	安全资格证书	safety qualification certificate
83	岗位技能证书	job skills certificate
84	事前控制	pre control
85	事中控制	in-process control
86	事后控制	post control
87	保质量	quality assurance
88	保安全	ensuring safety

续表

序号	中文	英文
89	保进度	ensure progress
90	安全帽	safety helmet
91	安全带	safety belt
92	安全网	safety net
93	开工令	work commencement order
94	停工令	work stoppage order
95	复工令	resumption of work order
96	项目业主	project owner
97	通信	communication
98	信号	signal
99	电力	power
100	施工安全	construction safety
101	施工进度	construction progress
102	投资控制	investment control
103	同吃	eating together
104	同住	cohabitation
105	同劳动	working together
106	同学习	learning together
107	同管理	managing together
108	营业执照	business license
109	税务登记证	tax registration certificate
110	组织机构代码	organizational code
111	资质证书	qualification certificate
112	银行开户许可证	bank account opening permit
113	施工围挡	construction fence
114	安全文化	safety culture
115	安全法制	security legal system
116	安全责任	safety responsibility
117	安全投入	security investment
118	安全科技	security technology
119	监督电话	complaint telephone number

续表

序号	中文	英文
120	文明施工	civilized construction
121	食堂	canteen
122	宿舍	dormitory
123	厕所	toilet
124	澡堂	bathhouse
125	医务室	medical room
126	娱乐部	recreation center
127	防冻	antifreeze
128	防触电	anti electric shock
129	防火	fireproof
130	防寒	cold prevention
131	防煤气中毒	prevent gas poisoning
132	防伤亡事故	casualty prevention
133	防车祸	preventing car accidents
134	保暖	warm
135	施工员	construction worker
136	材料员	materialist
137	测量员	surveyor
138	试验员	testers
139	资料员	data officer
140	安全员	security officer
141	质检员	quality inspector
142	监理工程师	supervision engineer
143	队长	captain
144	技术负责人	technical director
145	技术员	technicians
146	质量员	quality officer
147	安全员	security officer
148	领工员	leader
149	工班长	team leader

第三节 对话口译

Dialogue 1: Opening a Bank Account（音频：12.3.1）

Word Tips

活期账户	current account
定期账户	deposit account
储蓄账户	savings account
信用账户	credit account
支票账户	checking account
中国银行	Bank of China
转账	transfer
兑换	exchange
汇入汇款	inbound remittances
汇出汇款	outbound remittances
手机银行	mobile banking
网上银行	online banking
工作日	business day
银行职员	bank clerk

【场景：Tim is opening a bank account with the help of a bank clerk.

在银行职员的帮助下，Tim 正在开通银行开户。】

银行职员：早上好，欢迎来到中国银行。请问有什么可以帮您？

Tim: Good morning. I would like to open a bank account.

银行职员：没问题。您想开哪种账户呢？活期账户还是定期账户？

Customer: What's the difference?

银行职员：活期存款账户是储蓄存款的一种存款方式，可以办理现金存取，用于个人用途的转账、兑换、汇入和汇出汇款。活期账户可进行中国银行内部本人及他人活期账户之间的转账和兑换，可用于境内外汇款、入账。您也可开通网上银行和手机银行服务。另外，我行会给予一定的活期利息。定期存款账户是储蓄存款的一种定期存款方式，不可以办理现金存取，不能用于转账、汇入和汇出汇款。我行为客户提供七天、一个月、三个月、六个月、九个月、一年等不同期限的定期存款。定期存款到期后自动转存,您也可提前全额支取。定期账户可开通网上银行和手机银行服务,并在网上银行和手机银行端自行操作。开立定期账户的前提条件是开立活期或支票账户。相对于活期存款,定期存款利率较高。

Tim: I see, actually I want to apply for a credit account.

银行职员:没问题,请您填写一下这张申请表。

Tim: No problem.

银行职员:你想要多少信用额度?

Tim: I would like a 10,000 yuan spending limit.

银行职员:好的,我们会试着为您申请一下。

Tim: Wonderful. Will I also collect points when I use the card?

银行职员:是的,我行中银系列信用卡,长城系列信用卡的普卡、金卡、钛金卡产品,每消费1元人民币累积1分。

Tim: Perfect, I have filled out the form. Do you need anything else?

银行职员:请出示您的护照和工作证明。

Tim: Here you are.

银行职员:谢谢。您的信用卡申请已提交。如无特殊情况,我行将在 15 至 20 个工作日内,将卡片邮寄到您的地址。

Tim: Thank you for your help, have a good day.

银行职员:谢谢您! 祝您生活愉快!

Dialogue 2: Office Building Renting(音频:12.3.2)

Word Tips

租赁协议	lease agreement
条款	provisions
出租方	lessor
承租人	lessee
房屋	premises
整修	refurbish

正式发票	official invoice
税收收据	tax receipt
日历月	calendar month

【场景:Chen Dong, office manager of the China Road Bridge Corporation, is renting a four-storey office building on Cohen Street through Smith Estate Agency.

中国路桥公司办公室经理陈东正在通过史密斯房地产公司租用科恩街的一栋四层办公楼。】

Estate Agent: Good afternoon. Smith Estate Agency.

陈东:下午好,我是陈东,中国路桥公司办公室经理。我们想租科恩街上的一栋四层办公楼。

Estate Agent: Hello, Mr. Chen. Did you receive the draft of the lease agreement I sent to you the day before?

陈东:是的,我收到且读过了,现在打电话来是与您讨论一下条款。

Estate Agent: No problem. I'd like to explain the key provisions for you as you like.

陈东:非常感谢。你能详细解释一下出租方的义务吗?

Estate Agent: Sure. First of all, the lessor should provide the lessee with the premises and attached facilities on schedule for office using.

陈东:那么请在正式租赁协议中附上家具清单的详情。

Estate Agent: OK. And the lessor will pay the refurbishing and repairing expenses in case the premises and attached facilities are damaged due to quality problems, natural damages or other disasters.

陈东:我们希望出租方在收到租金的当天向承租人提供正式发票或税务收据。

Estate Agent: There is a specific provision here on what you said.

陈东:好的。接下来是付款条款。

Estate Agent: The lessee should pay the monthly rent by the twentieth of each calendar month.

陈东:可以。请问是否有关于如何合理使用房屋的规定?

Estate Agent: Yes. The lessee should not sublet the whole or any part of the premises.

陈东：明白了。办公室装修有什么规定吗？

Estate Agent: With the lessor's approval, the lessee can decorate the premises and add new facilities but should not change the good conditions of the premises for normal use.

陈东：好的。接下来是违约问题。

…………

第四节　篇章口译

Passage：Why construction sites need proactive security?（音频：12.4.1）

Word Tips

construction sites	建筑工地
proactive security	安全性
video surveillance	视频监控

【场景：A project manager is talking about why construction sites need proactive security and how to secure construction sites.

某项目经理正在谈论为什么及如何主动保障建筑工地的安全性。】

Construction sites are undoubtedly one of the most challenging environments to protect from a security perspective. Unlike the four walls of a fully completed structure, where workers and visitors can be closely monitored and funneled through designated entrances and exits, construction sites are a virtual free-for-all by comparison, with multiple points of entry and exits with various contractors and subcontractors coming and going at different times throughout the day.

Construction sites also feature a proverbial collection of highly desirable and easily resalable goods—ranging from lumber and copper that have skyrocketed in price with the supply chain crisis that has emerged across the USA to power tools and heavy machinery. In fact, according to the most recent Equipment Theft Report compiled by the National Equipment Register and National Insurance Crime Bureau, the estimated total value of stolen equipment in the U. S. alone in 2016 was nearly $300 million.

Traditionally, contractors have employed various tactics to mitigate against theft, such as installing perimeter fences and hiring guards to watch over property and ensure that only authorized personnel is allowed onsite. However, while patrolling guards may be a preferred method of preventing and deterring would-be thieves, they are not without drawbacks.

First and foremost, hiring in-house guards or outsourcing the function to a sub-contracted provider can be quite costly, especially at large sites where more than one person is needed to cover the property adequately. Second, guards are human and unable to keep tabs on every site area 24/7. Autonomous robots, or unmanned ground vehicles (UGVs) as they are commonly referred to, have also been presented as a solution to this dilemma. Still, they are costly and can only patrol for a finite period before they must be recharged.

The changing role of video surveillance

Video surveillance has also traditionally played a significant role in protecting construction sites, however, even if cameras are properly placed and maintained—a significant challenge within a construction zone—the footage captured is typically only used to document reported crimes or injuries after they have already occurred. Additionally, people are so accustomed to the ubiquity of cameras that they may fail to have the deterrent effect they once did.

The development of new video analytic tools powered by machine learning and artificial intelligence (AI) technology may help make surveillance a more proactive security tool for construction sites. For example, one of the biggest challenges historically in outdoor security applications has been the accidental activation of virtual tripwire on the perimeter of premises by things like animals or foliage blowing in the wind. However, video analytics can learn and differentiate between curious deer or falling tree limbs versus a person or vehicle coming onto the property. Some video analytics can also classify and track people, vehicles, and objects as they traverse a scene, providing security and first responders with an accurate description of potential trespassers.

Transitioning from reactive to proactive security

Even with video analytics, surveillance alone does not create a proactive security strategy. Combining surveillance and analytics with a live operator helps construction security professionals manage security threats.

Just because a guard is on duty does not mean they can respond to every potential intrusion, even when aided by advanced notice. It takes very little time for thieves to breach the perimeter of a building under development and pile thousands of dollars' worth of wire bundles, lumber, and tools into the bed of a pickup truck, or worse yet, drive off with heavy equipment that is not as readily available as it once was given the current supply chain crunch.

If individuals are captured on video entering a premise today, a live agent in a 24-hour monitoring center can be alerted to the breach and take a wide range of actions to address the situation, such as directly warning the intruders that they are being observed and recorded and that the authorities are being notified. In those circumstances where guards are also employed, the agent could dispatch them to the precise area of the intrusion and let them take over the response.

For years, video surveillance has primarily served as a post-incident investigation tool. Still, the advent of analytics and the evolution of live monitoring means that there are more strategies available to tackle the myriad risks that are present on today's construction sites. Moving from a reactive security posture to one that is proactive could mean the difference in completing a project that is on time and under budget as opposed to one that is delayed and overextended financially.

第五节　口译技能——工程口译中的跨文化意识

口译是指译员通过口头表达的形式将所获得的信息准确而又快速地由一种语言转换成另一种语言的过程，是人类在跨文化、跨民族交往活动中所依赖的一种基本的语言交际工具。但这种转换并非单纯的语言转换，而是文化信息的传递。从某种程度来说，口译所要翻译的不仅仅是交际双方不同的语言，还包括不同的文化。巴黎释意学派代表人物塞莱斯科维奇提出口译是一种交际行为，而非简单的语言行为，在话语所传达的意义与意义借以形成的语言之间存在着根本的差别。她认为，词语在语言层次具有多重含义，翻译不是单纯地从语言 A 到语言 B 的解码和编码，随着语境的扩展和认知知识补充的启动和参与，词语在语言层次意义上的多重性被单一的语篇意义替代。口译的目的是让听者明白说话人在语言层次之外的文化信息，从而达到思想上的交流。人们对文化的定义有诸多歧义，其中比较经典的是来自英国人类学家泰勒的定义。他指出文化是一个复杂的整体，包括实物、知识、信仰、艺术、道德、法律、风俗以及从社会上习得的能力与习惯等。语言是文化的一种表现形式，又是文化的重要载体。由于每一个民族对于自己的文化有不同的表达形式，因此各民族之间必然存在文化差异。文化差异的外在表现为语言表达、风俗习惯、体态表情等，而深层次的差异则体现在思维方式、价值标准、道德观念、审美观、心理因素、对世界的感知上。文化的社会性和民族性对口译造成了影响。译员在不了解文化差异的情况下进行的翻译，容易使听众产生困惑甚至误解，导致双方交际失败。

工程口译有其特殊性。工程口译的服务对象比较复杂，来自各种文化。在涉外工程项目口译中，译员光有语言知识及行业背景知识还不够，还必须有非常强的跨文化交际意识。此外，单就行业术语而言，不同的国家也有不同的习惯用词。在"一带一路"及中国企业"走出去"的大背景下，双方文化上的差异很有可能成为制约工程合作成功的因素。每个国家的文化背景，政治经济状况，词语的典故，风俗习惯，当地人民的民族信仰、价值观、审美观、禁忌，部分词语的联想意义等都不同。共建"一带一路"国家众多，涉及 7 大语系，使用的语言有 2400 多种，有着错综复杂的语言环境。如果不了解语言背后的文化内涵，就不能准确理解和表达出语言层次之外不同文化背景下交际双方的真实意图，由此可能会引起误会,甚至导致双方交际失败,造成工程项目面临经济损失。因此,译员要设法应对文化差异,培养跨文化交际意识。

(一)译前做好背景知识的储备

译员需要对当地的宗教信仰、饮食、服装、民俗习惯等有所了解,才能和客户聊得深,从而拉近关系。译员需要知道或者提前准备这些知识,方能胜任口译工作。所谓价值观是指人们认定事物、判定是非的一种思维或价值取向,各个国家和地区存在不同的价值观。西方社会比较崇尚个人主义,而亚洲一些国家则注重集体主义及等级观念,在口译中要注意差异,区别对待。例如:柬埔寨人等级观念较强,比较重视官爵、等级,所以在与柬埔寨人进行工程合作时,翻译要注意多用敬语,不要越层沟通;另外,有些中国人聊工作时有时也喜欢顺带谈论时政,但在与柬埔寨人聊工作时,就尽量不要谈及与政治相关的敏感话题,这是他们忌讳的话题。

(二)熟悉不同国家行业术语的习惯表达

虽然有些术语是通用的,但同样的概念在不同的国家和地区往往可能用不同的语言来表达,译员要尽量使用对方常用的专业用语。例如在电气化领域,因为中国最早引入的是德国技术,所以在说到"接触网"一词时,我国采用的是德国西门子常用的 OCS(Overhead Catenary System),而英国常用 OHL(Over Head Line);在谈到"隔离开关"时,英国人用"switch",德国人则喜欢用"disconnector"。又如"码头"一词,我方译员习惯使用"dock",而在孟加拉国,对方常用的是"jetty"。"路基"一词,中国人喜欢用"roadbed"或者是"subgrade",而柬埔寨人喜欢用"foundation"。在涉外工程项目的具体施工过程中,每个国家可能都有约定俗成的用词,这个词不一定非常准确,但因为使用频率特别高,已成为口译中的习惯表达。例如,由于中国建筑工程承包商在柬埔寨的项目较多,对于"级配碎石"(graded gravel,指颗粒大小有规定的碎石)的口译可以简化为汉语拼音"ji-pei",对当地柬埔寨的经销商或者施工人员说"ji-pei",他们也能明白是什么意思,后来他们也开始用这个词。施工现场讲究效率,对于术语的口译可以从简,对方能听明白就行。

(三)充分发挥主观能动性,灵活机动地翻译

译员还要注意化解文化差异带来的文化冲突。译员并非机械的"传声筒",在文化差异带来了误会时,要灵活机动地做些适当的调整,把说话人想要表达

的意图翻译出来。涉外工程的外籍人员可能来自世界不同的地方,饮食方面有不同的禁忌,如果遇到了有关对方禁忌的词语,一定要灵活处理,不能完全按原词翻译,可以省略不译或者用其他的词语翻译。译员需要发挥主观能动性,及时弥补说话人因为忽视文化差异而引发的失误,缓解可能由文化差异引起的矛盾和冲突,准确地把握、传递说话人的交际意图,从而帮助交际双方进行成功的跨文化交际。口译过程中需要具备跨文化交际的敏感性。

第六节　课后练习

（音频:12.6）

Ⅰ. Technical Terms Interpretation

1. 后勤保障

2. 仓库

3. 物资补给

4. 固定资产

5. 设备维护/检修

6. 财务计划

7. 会计核算

8. 车辆管理

9. 水电管理

10. 医疗保健管理

11. logistics management department

12. state-owned assets

13. energy management

14. infrastructure construction

15. maintenance and renovation

16. social service enterprises

17. green conservation

18. property services

19. water, electricity, and heating supply

20. dormitory management

21. medical care and public health

Ⅱ. Sentences Interpretation

1.接种疫苗通常取决于是否方便。为了帮助员工尽可能容易地接种疫苗,雇主可以支付儿童保育费或前往疫苗接种地的差旅费用,或者为员工提供带薪休假以促其接种疫苗。重要的是要给员工参加疫苗接种预约所需的时间和便利,直至其从疫苗接种中恢复。

2.解决人们对接种疫苗的担忧的最好方法之一是将他们推荐给他们信任的人。请记住,这些担忧可能涉及情感问题,也可能涉及科学之外的问题,例如他们的个人经历以及他们对过去不好的或不公平待遇的看法。不要用事实和信息压倒人们。相反,认可他们的观点或经历,使他们说出他们的动机,而不是一味输出你认为他们需要听到的信息。

3.物业管理的设施设备主要包含供电线路、各类配电设备、给排水管网、二次生活用水加压设备、电梯、消防设备等,小到宿舍区内无处不在的一盏路灯,大到几十万配电箱,这些设备加起来数量上千个。

4.内外网供电系统故障和宿舍内突发停电,应急供电系统启用管理和用电线路故障及时抢修以及供水管道的应急抢修,是后勤保障部门的所有工作。这些故障不管是发生在白天还是深夜,发生在严寒还是酷暑,后勤保障部门都责无旁贷,要第一时间投入抢修中去,宗旨是第一时间恢复和保障住户的正常生活秩序。

Ⅲ. Dialogue Interpretation

【场景:A Chinese engineer Peiling is planning to get vaccinated.
中国籍工程师佩玲准备在当地接种疫苗。】

Peiling: Vanita,怎么样了?

Vanita: Hey Peiling, have you booked your flu shot appointment yet?

Peiling: 是的,我刚刚预约了。我预约的是上午 10 点。你呢?

Vanita: I booked mine for 10:30 am. Since our timings are close, do you want to meet at the clinic and get together?

Peiling: 好的,我们9:45在诊所外见面吧。

Vanita: 9:45 am works for me. I'll see you there!

Peiling: 我们的预约时间略有不同,你觉得他们会允许我们一起打针吗?

Vanita: Ah, I'm not sure. But let's ask.

Vanita: Excuse me, my friend's appointment is at 10 and mine is at 10:30. Is it possible for us to get our flu shots together?

Nurse: Let me check. Can I have your names please?

Vanita: I'm Vanita, and this is Peiling.

Nurse: You can get your shots together. Just one person ahead of you, and then we'll vaccinate you both.

Vanita: Thank you.

Nurse: All set, who's going first?

Peiling: 我先,因为我的预约比较早。

Nurse: Which arm would you prefer, left or right?

Peiling: 左臂。

Nurse: And left or right for you?

Vanita: Right arm for me.

Nurse: There you go! Let me put a band-aid on that for you.

IV. Passage Interpretation

If a salary and bonus conversation is not going well, managers should spend more time listening to the employee in order to understand where they are coming from and what their concerns are.

There is often a lot to be gained by managers who are curious when it comes to having tough pay-related conversations. For example, they might learn that an employee feels that their job has not been correctly benchmarked against competitors or the wider market. This may or may not be true, and you should be willing to investigate the perceived discrepancy before making a final decision.

More often than not, a challenging conversation around salary and bonus will require a follow up meeting, giving managers an opportunity to come back with more facts and secure a positive outcome with the employee. Ultimately, the goal is to reach a mutually satisfactory decision where your employee feels valued and appropriately compensated by the company, and you are happy with the numbers and how the business's income and output balances.

Salary and bonus conversations should not be taken lightly, as they can have a tremendous impact on the productivity of your business: whether your team feel undervalued and become demotivated or decide to leave; or your employees are delighted with their new salary and their company advocacy has been given a boost. Put the time into getting these discussions right and your staff will soon be singing your praises to anyone who'll listen, helping you to attract new talented team members as well as new customers.

参 考 答 案

第一章　工程口译概述（参考答案）

第六节　口译技能——多任务处理

三、课后练习

This document stipulates the obligations and rights of the parties to the contract as well as the specific conditions for the execution of the project. In most cases, conditions of contract consist of two parts, that is, general conditions of contract and particular conditions of contract. The employer would designate a standard form of contract like FIDIC Conditions of Contract as his general conditions of contract, which will not be sent to the contractors with the invitation to tender. The name of the standard form and the version, however, will be stated in the instruction to bidders, or alternatively, the employer may provide his own version of conditions of contract if he chooses to do so. Amendments made to the general conditions of contract according to specific project conditions constitute the particular conditions of contract.

第二章　工程项目启动（参考答案）

第三节　对话口译

Dialogue 1:公司推介

Z:How do you do?

J:您好！

Z:I am Zhang Peng, project director of Jiangxi Water and Hydropower Construction Group Co. Ltd. Here is my business card.

J:谢谢！这是我的！

Z:Please allow me to give you a brief introduction to our company.

J:好的。

Z:This is our brochure.

J：照片很漂亮。

Z：Thank you! Our company has built many significant projects, and the brochure just presents some of them. By the way, our company was established in 1956.

J：那么,贵公司已经经营很多年了。

Z：Yes, we have made great achievements during the past 68 years.

J：贵集团是否只做建筑工程?

Z：Though famous for our hydropower station, we are also involved in other civil works, such as highways, bridges, pipeline engineering, etc. We are now looking for opportunities to take environmental protection projects.

J：你们有国外施工或同国际承包人合作的经验吗?

Z：Yes, we have. Our company was involved in construction projects in the East Africa, West Africa and Southeast Asia. We also had very smooth cooperation with contractors from France, Germany, Japan and other countries.

J：你们公司的注册资本大约是多少?

Z：About RMB 300 million Yuan.

J：很好,希望我们有机会合作!

Z：I am looking forward to cooperating with you and your company soon.

Dialogue 2：全球工程承包市场

Mr. Anderson：张鹏,我的老朋友,好久没见啦。你还好吗?

张鹏：Mr. Anderson! I'm fine, how about you?

Mr. Anderson：我也很好。我听说你已经被提升为贵集团副总裁啦。恭喜啊! 鉴于你在国际工程市场上的出色表现,我并不感到意外。

张鹏：Thanks for your compliment!

Mr. Anderson：过去几十年里,中国发展迅猛,中国建造承包商在国际市场的份额持续攀升,太神奇了!

张鹏：It is hard to believe that our last meeting was about ten years ago. It is more than 20 years since China joined the WTO. Since then Chinese contractors have grown rapidly in the international construction market. My company has also made great achievement under the guidance of going-global strategy made by the Chinese government. However, to be honest, my company is facing certain bottlenecks and problems, which I'd like to share with you. Taking this opportunity, may I be

245

honored to hear your foresight, and understand how you see future developments in the international engineering & contracting market?

Mr. Anderson:好啊,我也注意到国际工程承包市场的一些变化,很乐意跟你讨论。但是你为什么不先描述一下你公司目前的处境呢?

张鹏:Great! Our company's main business has been concentrated in underdeveloped countries in Asia, Africa, and Latin America. But encumbered by the fallout from the financial crisis in 2008 and European debt crisis in 2012, the economic activity of these countries has experienced a big down-turn and suffered deep slumps one after another. This has seriously affected the international trade and construction market. The underdeveloped countries that heavily rely on the export of natural resources for foreign exchange were struck heavily by the decline of crude oil and minerals prices. Accordingly, their fiscal revenue declined sharply and their construction budgets were cut down deeply. Sadly, some of our projects' payments were delayed severely. Recently we have noticed that new projects generally require contractors to have the ability to provide finance and investment.

Mr. Anderson:我也注意到了这个趋势。这些困难似乎不仅是中国承包商所面临的,也是世界各地其他承包商所都面临的。我认为,在亚洲、非洲、中东欧和拉丁美洲国家中,亚洲国家经济形势相对好于其他地区。

张鹏:Oh, for us the Middle East is also important customer, but the Middle East is volatile, and in many countries the situation is very tense and unpredictable, even in the short-term. For example, the war in Syria has been going on for years, and the Israeli-Palestinian conflict is not resolving itself.

Mr. Anderson:不错。但是如果仅仅考虑中亚、东南亚国家和印度,这些国家的经济势头正在复苏,显示出强劲的发展潜力。我相信他们政府有能力投资项目并及时付款。

张鹏:Great, we share the same perspective. These countries are adjacent to China and they have always been our key markets. But you know, competition in these countries is becoming more intense.

Mr. Anderson:我同意你的观点。在我看来,非洲幅员辽阔、资源丰富,是最具潜力的大陆。大多数非洲国家仍处于经济发展的初级阶段。所以他们专注于基本的基础设施建设和增强民生保障。

张鹏:Our company is working on many projects in Africa too, including some aid projects funded by Chinese government as well as more projects financed by Chinese banks.

Mr. Anderson:拉美市场也在快速增长,不过你要密切关注那里的市场风险。

张鹏:You are right! Due to the lower price of bulk commodities, many developing countries relying on natural resources have encountered tough circumstances. Because of the sharp reduction in capital investment budgets and the ability to provide reliable financial guarantees, F + EPC projects are less and less adopted, while the PPP, BOT construction projects are more and more common. In this new market environment, our company is trying to transit its business model to adapt to the current situation.

Mr. Anderson:中国政府以其远见卓识提出了"一带一路"倡议,为推动世界经济特别是为国际工程承包市场创造机会发挥了重要作用。中国还牵头创建了亚洲基础设施投资银行,支持发展中国家的基础设施建设,我对此十分钦佩。

张鹏:The initiative proposed by Chinese government does play an important role in promoting the international contracting industry. Our company works closely with the big policy banks such as the Export-Import Bank of China, the China Development Bank and other state-owned commercial banks like the Industrial and Commercial Bank of China, and the Bank of China to assist foreign governments in financing and promoting appropriate infrastructure projects. At the same time, we also have good links with various funds, such as the Silk Road Fund, the China-Africa Development Fund and the China-Africa Fund for Industrial Cooperation to implement investment projects, PPP and BOT projects.

Mr. Anderson:这真是非常鼓舞人心啊。事实上,发达国家也面临着基础设施建设资金不足的问题。我希望你们能来我的国家进行调研,并以 PPP 模式开展业务。我会把我的一些好朋友介绍给你,他们将对你们的业务初创有很大帮助。

张鹏:Thank you very much, Mr. Anderson. I would be very glad to cooperate with you in the future.

第四节　篇章口译

Passage 1:英译中

大家早上好! 首先请允许我做自我介绍,我是本项目的施工经理张鹏,负责本项目有关施工的具体事宜。现在,我想跟大家简单介绍一下我们的施工计划。我们一定要全力以赴,做好施工前准备工作,并做好各项施工方案的审定工作。

施工组织计划(COP)是一份详细而全面的文件,概述了施工项目的规划和执行。COP 的主要目的是在对周围环境干扰最小的情况下确保施工工作顺利、安全、高效。准备充分的 COP 有助于降低风险、控制成本并按时交付项目。

步骤1:现场检查和评估

编制 COP 的第一步是进行彻底的现场检查和评估。本次检查的目的是识别可能影响施工的所有潜在危险或障碍。评估应包括土壤条件、现有结构、地形、通道、排水系统和公用设施等因素。

步骤2:确定项目团队

一旦现场评估完成,下一步就是确定项目团队。该小组将负责制定 COP 并确保其正确实施。项目团队应由具有相关专业知识和经验的人员组成。该团队应包括一名项目经理、施工经理、工程师、安全员和其他关键人员。

步骤3:制定 COP

下一步是制定 COP。COP 应包括关于项目范围、进度、预算、施工方法、质量控制、安全程序和风险管理的详细信息。COP 还应概述项目团队和参与施工工作的任何分包商的角色和责任。

步骤4:与利益相关者协商

在制定 COP 时,与利益攸关方进行协商是很重要的。这包括地方当局、监管机构和社会群体组织。与利益相关者协商有助于确定与施工工程相关的任何问题或影响。可以将这些信息纳入 COP,以确保该项目以对社会负责和可持续的方式进行。

步骤5:审查和调整 COP

COP 一旦完成,就应进行必要的审查和调整。审查的目的是确保文件全面有效。审查应考虑项目范围、现场条件或法规的变化等因素。可能需要对 COP 进行调整,以确保项目安全、高效地实施,并符合要求的质量标准。

总之,施工组织计划(COP)是任何施工项目的重要组成部分。它使项目团队能够以安全、高效和可持续的方式规划和执行施工工作。COP 的制定应采用结构化的方法,包括全面的现场评估、关键人员的确定、与利益攸关方协商以及定期审查和调整。

Passage 2:中译英

Good morning! I'm Zhang Peng, working for Jiangxi Water and Hydropower Construction Group limited. I'm the civil contractor representative. First of all, thank you for giving my company this opportunity. Next, I would like to introduce my company to you. Our company is one of the top 250 largest international contractors in the world and one of the top ten leading enterprises of "going global" in Jiangxi province and we have rich experience in hydropower station construction, highway engineering, building engineering and electromechanical engineering construction.

In the international market, our company, as one of the construction enterprises with earlier participation in international business and higher degree of internationalization, has continuously accelerated the pace of "going global" and vigorously promoted the "Overseas Strategy". The group has set up branches and offices in more than ten countries, including Kenya, Ethiopia, Mozambique, Uganda, Benin, etc., and actively carries out foreign investment and construction projects, with strong comprehensive operation and management capabilities in new modes such as PPP (Public-Private Partnership) and EPC (Engineering Procurement Construction).

Since 1993, the company has been involved in overseas and participated in international market competition, and has constructed the water conservancy projects of Hali River, Erbaye and Gelila, the asphalt core wall project and tunnel project of WOLKITE Dam, the Ribb Irrigation Project in Ethiopia, the Twin Rivers Shopping Mall Project, three small hydropower stations, the Nairobi Power Grid Upgrading Project, the Victoria Bank Project, KENHA Office Building Project, Alma Apartment Building Project in Kenya, Nam Hwa Hydroelectric Power Station in Vietnam, Highway Project in Mongolia, Bindura Dam Project in Zimbabwe, Two Dams Project in Tunisia, Construction of Healthcare Center in the Kingdom of Tonga, Parwan Hydraulic Project in Afghanistan, Pacat Hydroelectric Power Station Project in Indonesia, Corumana Dam in Mozambique Project, Uganda Defense Hospital Project, Rwanda Rusumo Hydropower Station Project, Rwanda MIG Commercial Complex Project and nearly 100 other projects.

In recent years, with the proposal of the Belt and Road Initiative, the company has gradually diversified the ways of exploring overseas markets and participating in overseas construction through the implementation of the "Overseas Strategy", with a steady increase in the undertakings and output value. In terms of market development, the company has changed from the project-driven market mode to the proactive layout of overseas markets in combination with the "Belt and Road" initiative and various policies. By basing itself on East Africa, the company radiates to the whole of Africa. At the same time, the company has increased market development in countries along the Silk Road Economic Belt, such as Pakistan, Myanmar, Uzbekistan, Kazakhstan, Georgia and other countries. In addition, the company is also stabilizing the Southeast Asian market with Indonesia as the base and diversifying its business growth through foreign investment.

We will attach some brochures with detailed information about our company to the prequalification letter. Through these brochures, you will learn more about the organizational structure of our headquarters, subsidiaries and affiliates, as well as the construction machinery and equipment owned by our company.

第五节　口译技能——主旨听辨

三、课后练习

Conflicts between the contractor's pursuit of profits and the employer's maximization of benefits are inevitable.

第六节　课后练习

Ⅰ. Technical Terms Interpretation

1. international contractor

2. annual turnover

3. international engineering & contracting market

4. sovereign debts crisis

5. drawing

6. blue print

7. the Belt and Road Initiative

8. deploy

9. site inspection

10. Public Works Bureau

11. 融资加设计—采购—施工模式

12. 项目总监

13. 经济势头

14. 地形

15. 丝路基金

16. 土木工程

17. 管道工程

18. 施工计划

19. 通道

20. 结构化方式

Ⅱ. **Sentences Interpretation**

1. It might be better for you to have a rest. We will go to the site tomorrow.

2. Though famous for our hydropower station, we are also involved in other civil works, such as highways, bridges, pipeline engineering, etc.

3. We are now looking for opportunities to take environmental protection projects.

4. Our company is one of the largest construction companies in China, with very rich experience in the construction of coal-fired power stations, nuclear power stations and hydroelectric power stations.

5. As a general contractor, our company just finished building a 2×600 MW coal-fired power plant in Sichuan Province last year.

Ⅲ. **Dialogue Interpretation**

L:Excuse me! Are you Mr. Williams?

N:是的,我就是。我叫尼尔·威廉姆斯,就叫我尼尔吧!

L:How do you do, Mr. Williams? My name is Li Xiang, the assistant of project director.

N:您好,很高兴见到您,李想先生。

L:You know, project director Mr. Zhang asked me to pick you up to your hotel.

N:谢谢!

L:How was your trip?

N:不是很好,我有点晕机。

L:Do you feel better now?

N:好点了,但是我觉得有点累,因为坐了很长时间飞机。

L:OK, here we go. Let's get you checked in at the hotel. You may have a good rest today and we'll go to the project site tomorrow.

N:好的,谢谢您!

L:Please wait for me here. I'll go the parking lot to get my car.

N:好的,没问题!

IV. Passage Interpretation

建筑材料是任何建筑的基本要素。按照用途,建筑材料可以分为三种。结构材料是指那些支撑建筑物的,使建筑物保持刚度的,形成墙壁和屋顶这样的外表面覆盖物的,将建筑物内部空间分隔成房间的建筑材料。第二种是建筑物内部的设备材料,例如管道、供暖和照明系统。最后一种是用来保护或者装饰其他材料的材料。

第三章　工程项目管理之招标投标管理(参考答案)

第三节　对话口译

Dialogue 1

Zhang Tian:嗨,陈先生。我叫张天,一个月前被人力资源部录用了。

陈亮:Hello, Mr. Zhang. Nice to meet you.

Zhang Tian:见到你也很高兴。我知道你是西非地区的项目经理。

陈亮:Yes, it is indeed so.

Zhang Tian:作为一个新人,我很想知道一些关于招标的事情,因为我想尽快进入我的岗位。你能抽出几分钟时间给我吗?

陈亮:My honor. What do you want to know?

Zhang Tian:据我所知,投标邀请书的一般信息和要求中有许多注意事项。它们是什么?

陈亮:Well, usually, they are: 1) The bidding price must remain stable within the specified number of days after the bid opening; 2) Regarding accidental closure; 3) The right to cancel or reject bids in your jurisdiction.

Zhang Tian:我明白了。开标后,投标价格应保持多久?

陈亮:It depends on the situation. For example, the IFB we will release next Monday informs bidders that the bid price should remain firm for 35 days.

Zhang Tian:好的,谢谢你的解释。我想有人在给你打电话,我下次会再来的。

陈亮：You're welcome. You can contact me on Tuesday, Wednesday, and Thursday.

Zhang Tian：非常感谢,再见。

陈亮：Goodbye.

Dialogue 2

Zhang Tian：这顿饭很好,我现在吃饱了。

赵康：Well, it seems that you have a good appetite. Let's continue, I know you have many questions.

Zhang Tian：你说得对。我还有几个关于IFB的问题要问。

赵康：Don't hesitate, may I ask!

Zhang Tian：好的。一定要举行标前会议吗?

赵康：Usually, for a large and complex procurement, a pre bidding meeting is held.

Zhang Tian：投标保证金有必要吗?

赵康：No, it's not.

Zhang Tian：关于定价和定价形式,有什么特别的吗?

赵康：It can be said that bidders must submit price information, such as unit price or hourly price and total price.

Zhang Tian：我明白了。规定了单位或小时费率和总价,有人告诉我,有些合同需要高度逐项化的任务描述。你能给我举个例子吗?

赵康：No problem. For example, IFB used for floor renovation may require the removal of gum, tar, and similar substances from the floor surface.

Zhang Tian：明白了。因此,高度逐项化的任务描述也可以是这样的:应该使用带磨砂刷的电动抛光机来获得更好的镜面光洁度。

赵康：That's right, your learning ability is very strong.

Zhang Tian：别太夸我啦!

第四节　篇章口译

Passage 1

首先,我要祝贺大家,在我们审查了你们的资格预审文件后,你们的公司已被接受进入发电站项目的下一阶段招标。这是土木工程的招标文件,你们会发现有很多卷。

第 1 卷为合同条款,根据 FIDIC 条款编制。你们知道,FIDIC 条款对承包商和客户都是公平的,所以我们希望你们能同意这些条款。

第 2 卷为工程量清单。本卷共有 18 节,费率 989 项。请记住,这里的合同价格是基于费率的合同,这意味着合同的总金额将取决于承包商实施的实际工程量。电站结构比较复杂,工程量较大。因此,在合同期内,每项工程的费率是固定的,但工程量将相应地进行修改。

第 3 卷为工程范围。你们能找到发电站的每栋建筑和主要结构的描述。你们应根据本卷中的总指导时间表提供每栋建筑的详细施工时间表。

第 4 卷为技术规范。电站与普通建筑大不相同,需要许多特定的标准和方法。由于所有有资格进入下一阶段的公司都是经验丰富的发电站承包商,你们不应该对本文件涉及的这些复杂的规范感到奇怪。

第 5 卷是质量保证。本项目定义了 3 个质量保证等级。

第 6 卷为图纸,所有图纸均加盖"仅供投标使用"印章。

招标文件第 7 卷包括协议书、施工方案、质量安全计划和工程量清单。

我们将根据对贵方投标文件的综合分析和判断做出决定。我们希望你们尽最大努力在 30 个日历日内完成招标文件的所有文件并将其移交给我们。祝你们好运!

Passage 2

如前所述,招标和投标是贸易交易的两个方面。事实上,买方和投标人的角色是可以互换的。也就是说,在各种情况下,购买者可能成为投标人,投标人可能变成购买者。

由于 IFB 对投标人提出了严格的要求,一个成功的投标方案肯定需要时间、知识、技能和努力。任何有陷阱的投标方案都不可能具有竞争力,即使是最微小的错误也可能导致错过一个有利可图的项目。以下步骤可能有助于准备中标方案和做好投标提交工作。

不同的项目有不同的要求。从实际的角度来看,你不可能对你遇到的每个项目都出价。赢得一份你的公司无法充分履行的合同可能与没有赢得合同代价一样大。因此,一旦你意识到如果你被授予合同,你的公司将无法获得合理的利润,你最好放弃它,继续下一份合同。

一旦你决定提交投标书,你需要运用你的技能,尽可能了解发布 IFB 的公司的一切情况,如其市场、组织、团队和其他事务。之后,您可以尝试联系潜在买家,以获得一些其他相关见解。当然,你需要访问网站并参加标前会议(如果有的话),因为这些活动可能会帮助你更好地投标。

如果你想赢得投标,你不能低估投标的重要性。如果你想让你的投标引人注目,你需要组建一个专业团队,在 IFB 所有规定和要求的基础上精心准备投标书。你应该严格遵守招标说明和回复表,这些说明和表格设置了严格的字数限制,并提出了旨在寻求理想回复者的问题。

投标书是如此复杂,以至于你需要在提交之前重新检查一遍。这些因素,如你的报价、参考资料、签名、拼写、语法、设计甚至美学,都非常重要,需要检查和确认。如果可能的话,你可以求助于一些值得信赖的朋友来评估你的投标书。一旦你的标书中出了一些错误,你要让你的团队成员及时修改。

不要拖延投标。发布招标的公司或组织极有可能拒绝任何逾期投标。考虑所有因素,如交货、计算机和互联网问题,并确保您的投标书能够及时送达合适的人或部门。如果允许的话,你甚至可以考虑亲自递交投标书。

总之,准备投标并非易事。如果你想赢得一份合同,以上五个步骤可能会对你有所帮助。

第五节　口译技能——信息分层

三、示例精讲

参考概述如下:

What is the role of government in new technology development?

From the traditional point of view, there are two main reasons for government's support. First, companies may invest less in research and development because they cannot always capture the benefits from doing research without the need in society. Second, the government needs to intervene in companies' economic decision due to environmental externalities.

From what has been found in environmental innovation study, there are also two reasons for this. One is that the government can help companies develop necessary skills and build more technology capacity. The other is that institutions are locked into particular patterns of incentives and behaviors, which need the government to change.

第六节　课后练习

Ⅰ. Technical Terms Interpretation

1. bidding documents

2. competitive bidding

3. notice of winning the bid

4. currency requirements

5. site visits

6. credit record

7. building construction contracts

8. private bank account

9. bid guarantees

10. performance guarantees

11. CPA（Certified Public Accountant）

12. publicly-funded construction projects

13. 授予合同

14. 投标邀请函

15. 原标书

16. 投标书有效期

17. 招标书附录

18. 投标提交日期

19. 专业分包商

20. 工艺标准

21. 补充附图

22. 通用规定

23. 法律诉讼

24. 仲裁

25. 经签字确认真实性的复印件

26. 资产负债表

27. 损益表

Ⅱ. Sentences Interpretation

1. Tender bond is designed to protect tenderee and buyer from losses because of any actions of tenderers. In case tenderee and buyer suffers any losses because of any actions of tenderers, the tender bond will be confiscated according to Clause 15. 7 of the Notice. We are deeply grateful for your official invitation to participate in the bidding for the aforementioned engineering project.

2. They undertake to submit tenders by the date required by the Notice to Bidders.

3. We have received all the tender documents and are preparing the bid.

4. Bidders shall submit a bid bond in the amount of not less than 2% of the total bid price as part of their bid.

5. Unfortunately, due to the heavy task of our project, we cannot participate in the bidding this time, please forgive me.

Ⅲ. Dialogue Interpretation

A：Excuse me, may I collect a tender document here?

B：您的公司通过资格预审了吗？

A：Yes, we were informed by your office two days ago.

B：您是哪家公司？

A：China Construction Group.

B：请出示你们公司领取招标文件的授权书和您的名片。

A：Here you are.

B：行,好的。招标文件包括3个分册和36张招标图纸。这是第一分册,包括投标须知、投标书、协议书和通用(专用)合同条件。

A：What are the requirements for the Performance Bond?

B：占合同总价的10%。有关合同条件的主要内容,例如工期,误期损害赔偿费,滞留金百分比和保修期等都已摘录出来,放在投标书附录 A 中。履约保函的格式在投标书附录 B 中。

A：I see. What about Booklets 2 and 3?

B：第二分册包括所有的技术说明书。

A：Is the Finishing Schedule also included in Booklet 2?

B：是的。您可在建筑工程一章中找到。工程量计算方法和工程量清单在第三分册中。这是投标图纸清单。您可对照此表检查一下图纸。

A：Thirty-six sheets of tender drawings, exactly.

B：请在这儿签字,以便我留底。

Ⅳ. Passage Interpretation

Hello everyone, please note! Please keep the following requirements in mind when preparing your bid:

（1）The bidder shall prepare one original and two copies of the tender in accordance with the regulations, and indicate "original" and "copy" respectively. If there is any discrepancy between the two, the original shall prevail. Then, the bidder shall put the original and two copies of the tender into an inner envelope and an outer envelope respectively, and indicate "original" and "copy" on the envelope.

(2) The original and two copies of the tender shall be printed or written in indelible ink and signed by one or more persons who have the authority to bring the bidder into compliance with the contract. In addition, there should be a written power of attorney evidencing the authorization along with the tender. Each page of the tender with entries and corrections must be signed by the tender signatory.

(3) There shall be no alterations, between the lines or smearing of the full set of tenders, except as directed by the agenda or for the correction of the bidder's error, in which case the correction shall be signed by the signatory of the tender.

(4) Each bidder can only submit one bid, and the bidder can only submit one bid for a contract.

第四章 工程合同谈判(参考答案)

第三节 对话口译

Dialogue:采购物资

Niko:Nancy 女士,我们已经审阅了你们起草的供油合同草案,有一些条款需要澄清或修正。

Nancy:So, can we discuss it one by one? We can see whether we can decide on it today.

Niko:好。咱们就从第二条"合同标的"开始。除了高级汽油和柴油,你们还能为我们供应各类非燃料用油吗? 你知道,我们的设备运行同样需要很多这类油。

Nancy:Esso company is happy to supply all kinds of non-fuel oils, such as oil, hydraulic oil and lubricating oil.

Niko:很好。现在我们看第四条,是关于供货合同期的。此条规定:"供货合同从本日起有效期为三年,并于 2026 年 1 月 1 日自动失效。"的确,根据我们现在的施工计划,这个高速公路项目的竣工期是三年,然而,项目可能会延期一段时间,也可能比计划提前(竣工)。这个谁也说不定,即使像我们这样有经验的承包商。我觉得可以把措辞改成这样:"供货合同期从 2023 年 1 月 1 日开始,并于高速公路项目竣工之日止。"

Nancy:I agree with you.

Niko:在第六条的第三段,也就是规定交货方式的那一条,其中的"买方的订单"是否包括口头和书面订单?

Nancy:Our intention is that "the buyer's order" is in writing and does not include an oral order.

Niko:我希望你们也接受口头订单。有时,我们急需某些油类,来不及给你们发出书面订单。

Nancy:Mr. Niko, we are willing to do our best to provide you with convenience, but oral orders are easy to go wrong.

Niko:我看最好这样写:"当收到买方的书面订单或口头订单(此种情况下随后要有书面确认函)……"

Nancy:I can accept that.

第四节　篇章口译

Passage 1:业主代表就合同细节发言

我们注意到施工机械费总计为801万美元。这个费用真的太高了！我们对常用机械折旧年限也有所了解。一般而言,大型耐用设备的折旧年限为15年,一般设备为10年,易损机械为3年。如你们所说,这个工程的工期非常紧张,必须采用施工设备清单中的许多高效率的设备,比如混凝土泵、大容量的混凝土搅拌站、混凝土搅拌车和许多种类的吊车等。

我们完全理解你们应当调动足够的机械设备来施工,但为了控制机械费用,你们最好准备一份与施工计划表相匹配的机械设备进场计划表来避免施工现场的机械设备闲置,并且检查一下机械设备费用,看看是否可以削减费用。

另外,工人和管理人员的营地费用太高了。你们临时建筑的预算为60美元每平方米,这完全没有必要,因为当地工人可以使用只需35美元每平方米的用当地的一种植物盖的房屋。

因此,我们需要对营地费用拟出详细的成本细分清单,可以分成两部分:管理人员营地的费用和工人营地的费用。我认为管理人员的生活区费用不会超过60美元每平方米,而工人的生活区费用也不会超过40美元每平方米。这样就可以削减一大笔费用。

第6项费用为主要分包商的费用。你们提及了预应力工程、墙围护和屋面工程、暖通空调工程的分包商。但是,我们不能接受这些专业工程的费用以几个包干价的形式呈交给我们。你们应当给我们提供像工程量清单中混凝土、钢筋、模板和其他项目那样的详细分析,否则这个价格细分清单就不完整。

Passage 2:承包人代表发言

We have the following comments on the conditions of this contract that we would like you to consider:

Clause 30 requires that the Contractor must commence the Works within 10 days after signing the Contract. We think it is a little bit too tight. This is a big project, and we need longer time to make preparations. Therefore, we suggest that "10 days" be modified to "15 days".

Clause 36 only mentions what the Contractor should do but does not mention what the Contractor is entitled to do in case the site is not handed over to him at the time agreed by both sides. Therefore, we think it is necessary to add a sentence like this: If the Contractor suffers delay or incurs costs for possession of the site or a part of the site, he has the right to obtain an extension of time to complete the Works and an amount of cost added to the Contract Price.

We can't agree to the liquidated damages for delay mentioned in Clause 43. First of all, the amount of the liquidated damages for each grade of key dates is too high. For example, you defined the amount of four-star key dates will be USD 20, 000 for each day. This is much higher than normal. We have completed the same kind of project last year. The payment for the liquidated damages in the contract conditions is only half of this one. Besides, since there is a penalty for delay of completion, we should demand a bonus for any early completion before the key dates.

We have a different opinion for the retention period and retention money in Clause 50. It is mentioned in the Clause that the retention period will be one year from the issuing of completion certificate of the whole Works. We think it is reasonable to calculate the retention period from the day of issuing completion certificate for each section, because there are long periods of time between sections in completion.

第六节　课后练习

Ⅰ. Technical Terms Interpretation

1. contract conditions

2. in line with

3. advance payment

4. contract value

5. progress payment

6. mobilization cost

7. performance bond

8. payment provisions

9. terms of payment

10. valid contract

11. 无效合同

12. 合同暂停

13. 合同终止

14. 履行义务

15. 履行合同

16. 违反合同

17. 转包,分包

18. 承包商,承包人

19. 承包商(人)代理人

20. 承包商人员

Ⅱ. Sentences Interpretation

1. We suggest that a sentence be added after the first sentence in Sub-Clause 6.6.

2. The contract contains basically all we have agreed upon during our negotiations.

3. Then what are the main points concerned in the Conditions of the Contract?

4. Once you win the tender for the project, you will receive the drawings stamped with "for Construction only".

5. This is a serious breach of contract for which we will require appropriate compensation and we reserve all our rights and remedies in this matter.

6. In case one party fails to carry out the contract, the other party is entitled to cancel the contract.

7. Are you worrying about the non-execution of the contract and non-payment on our part?

8. Please sign a copy of our Sales Contract No. 156 enclosed here in duplicate and return to us for our file.

Ⅲ. Dialogue Interpretation

D:今天我们将进入合同谈判的阶段。在这个阶段,我们将讨论合同条件和价格。根据时间表,我们在每个话题上要花几天时间。

Z:I am very glad to have the opportunity to discuss the terms of the Contract with you.

D:现在我们直击重点。首先,我想明确的是,合同条件符合 FIDIC 的一般原则。

Z:We are very in favor of FIDIC because it is fair to both the owner and the contractor.

D:我相信您一定已经仔细阅读了这些条件,现在请告诉我们您的评论。

Z:Frankly, if you don't mind it, we do have different views on this contract clause. There is no provision in the Contract for the advances paid to the contractor. Usually, after signing the Contract, the contractor should receive about 10% of the contract price to prepare for the project.

D:恐怕在这一点上已经没有讨论的余地了。根据合同规定,付款只能从进度付款开始,您自己必须为动员费用准备资金。

Z:In that case, we withdraw this opinion. Clause 8 is about performance security, and we think the 15% margin is higher than usual. We want to adjust it to 10%.

D:我们将考虑你对这个条款的建议。

Ⅳ. Passage Interpretation

尊敬的副部长,我很荣幸今天能与您见面。我们有下列五个建议。首先,我们认为一些重要信息应作为附件编入特许经营权协议,诸如项目现场位置、现场交付计划、技术要求、运维要求、移交程序、融资结构、资金流量预测以及政府支持协议等。其次,我们建议把 5 年施工期和 25 年的运营维保期简单合并为 30 年特许经营期。再次,我们建议特许经营权协议应明确禁止与之竞争的道路的建设,包括公路、铁路或者与之平行的道路的升级,那会分流项目的收入来源。然后,特许经营权协议应给出针对交通流率的补偿机制。最后,我们建议与消费者物价指数挂钩的过路费率实现指标化,相应地,根据贵国的通胀情况每年调整过路费。

第五章 工程施工准备(参考答案)

第三节 对话口译

Dialogue 1:讨论施工现场布置图

阿姆斯特朗先生:您好,白先生!

Mr. Bai:Hello, Mr. Armstrong!

阿姆斯特朗先生:现场布置图带来了吗?

Mr. Bai:Yes. I worked on the proposal yesterday evening up to midnight. I knew you were very concerned about it.

阿姆斯特朗先生:你知道同时在现场施工的队伍有很多,而整个场地很有限。你们应该把场地安排得紧凑一些。

Mr. Bai:I understand. Here you are! Precast yard, rebar yard, concrete batching plant, repair workshop, warehouse, and iron shop are all included.

阿姆斯特朗先生:哦,很详尽。根据你们提交的施工方案,预制混凝土桩将在现场制作。

Mr. Bai:Yes. Two gantry cranes shall be set here for handling the piles.

阿姆斯特朗先生:龙门吊的承载力有多大?

Mr. Bai:10 tons each.

阿姆斯特朗先生:在平面图中你们似乎有 8000 平方米的土地用作预制构件厂,这样,给其他设施的地方就不够了。

Mr. Bai:We will use that piece of land for the precast yard for a short period only. Piling works goes on for only 3 months, we will reduce the area for the precast yard and turn it into carpentry shop, warehouse, repair workshop, and iron shop. We will build a rebar yard for use during the structure construction period. The other facilities can be set up later.

阿姆斯特朗先生:钢筋车间包含在钢筋场内吗?

Mr. Bai:Yes. We will have a cutting machine and a bending machine in the rebar shop that can process rebar with a maximum diameter of 40 millimeters.

阿姆斯特朗先生:很好。现场建几个混凝土搅拌站?

Mr. Bai:Two. One with a capacity of 40 cubic meters per hour, and the other 25 cubic meters per hour.

阿姆斯特朗先生:混凝土搅拌站产生的废水将直接排入市政排水系统吗?

Mr. Bai: Of course not. It will be filtered through a catch basin before it drains down into the city drainage system.

阿姆斯特朗先生:我找不到你们的现场试验室。因为试验室是最基本的质量控制设施,平面图上可不能忽略了它。

Mr. Bai: It will certainly not be omitted. You see, that block not far away from the batching plant is the lab. It will be equipped with all required instruments for soil, concrete and steel tests necessary for the works.

阿姆斯特朗先生:现场办公室在哪?

Mr. Bai: Here it is. The office for our company can maintain 40 people working together, while the office for the client can maintain 15 people.

阿姆斯特朗先生:还有没有足够的空间来安排一个会议室?

Mr. Bai: Oh, I'm awfully sorry. We neglected it. Would you mind adding a 40 square meters area for it just beside your site office?

阿姆斯特朗先生:当然不会。这样我们交流起来就更方便了。

Mr. Bai: We will correct this layout immediately and resubmit it next week.

阿姆斯特朗先生:好吧。

Dialogue 2:施工人员招录

Mr. Fang: Welcome Mr. Jackson and Mrs. Morris. Thank you so much for your kind presence to our office on the first day of our local workers recruitment.

Jackson 先生:今天对贵公司和我们部门来说都是重要的日子。这个省成千上万的年轻人找工作已经找了很长时间啦。我怎么能不来呢?

Morris 夫人:我也是。我很高兴数百名年轻人将在这里找到工作,但与此同时,我也非常关注传染病控制的巨大挑战。

Mr. Fang: Mr. Jackson, I hope you can help us to solve the problem we are facing in local worker's recruitment. Some of the local candidates even don't have their ID card.

Jackson 先生:哦,在这个国家这可不是稀罕事儿。身份证还没有覆盖到一些偏远的农村地区。别担心,方先生,我会叫警察局长派一名警官过来给他们签发身份证。

Mr. Fang: You've done me a great favor. Thank you.

Jackson 先生:到目前为止,已经招募了多少本地工人了?

Mr. Fang: Our company is going to recruit 500 local workers this time, and up to now we have received more than 700 applications for employment already. There is a very strict recruitment procedure including several steps, i. e. application, ID authentication, health examination and interview. We can only confirm the acceptance of a local worker until he has positively passed all the steps. Since we will have the Health Examination Reports three days later, I cannot confirm the number of the recruited local workers yet.

Jackson 先生:三天？时间好长啊。本地的求职申请者在做什么样的体检啊？

Mr. Fang:Except general health check, we requested mainly two sets of health examination items. The first set for ascertaining if the candidate has any disease not suitable to work in the construction site, such as heart trouble, hypertension and physical disability, etc.; and the second set for eliminating any patients with infectious disease including AIDS, cholera, hepatitis, etc. The Health Examination Scheme was made under the advice from the Provincial Public Health Department. A lot of thanks to Mrs. Morris.

Morris 夫人:不用谢,这是我的职责所在。因为你们要在营地里聚集数百名工人,我希望和 Jackson 先生一起检查一下本地工人的住宿条件。

Jackson 先生:好主意。方先生,我们现在就去你的工人营地吧。

Mr. Fang:Glad to receive your inspection. That couldn't be better. The front is just the camp for local workers. We built it last month.

Jackson 先生:啊哈,这些房子都是用集装箱搭建的,不是吗？

Mr. Fang:Yes, used 150 containers for this camp, 125 for dormitories, and 25 for canteens, shower rooms, laundries, toilets.

Morris 夫人:真是令人印象深刻啊。你们给外立面刷了可爱的颜色,还在每个院子里种了树。我想看看房间里的设施。

Mr. Fang:This is one typical dormitory transformed from one container. We cut the door and window openings, made the interior wall finishing with thermal insulation, installed wooden floor and equipped air conditioner for the dormitory.

Morris 夫人:有四张床、一张桌子和一个书架。为什么在这里我找不到蚊帐呢？这里可是真正的疟疾高发区啊。

Mr. Fang:We've noticed that and purchased enough mosquito netting sets already. Every local worker will have one set as soon as he has been formally hired.

Morris 夫人：那我们检查一下淋浴间和厕所吧。

Mr. Fang：We built ten shower rooms with five showers each. You see, there are three solar water heaters above the roof of the shower room. The floor of shower room is of anti-slip ceramic tile, actually, even the canteens and toilets as well.

Jackson 先生：你们为本地工人建了一个很好的营地。三天后当你完成当地工人招聘时，我会再来。

第四节　篇章口译

Passage 1：如何进行施工准备

施工准备的定义是为在未开发土地上修建建筑物、停车场、道路等做必要的准备。施工准备是开始建筑施工前要做的初步工作之一，被选中用于施工的建筑工地需要进行适当的准备。

施工准备应考虑下列因素：

1. 施工准备的第一步是，如果现场有灌木丛或丛林，应先行清除。

2. 大致整平整个区域。

3. 将施工现场的洞穴填满沙子或夯土，并按权威机构的要求整平或重新定位。

4. 作为施工准备的一部分，树木将被砍伐，并按权威部门要求连根拔起。

5. 施工准备之前，必须在施工现场的适当位置建立永久性的测定基准点。

6. 建筑物的定向和壕沟应该正确设置在建筑工地，材料存储和堆放的位置应该明确设置在现场地面上。

施工准备包括几个不同的任务，例如沉积物控制、清理地面和挖掘树根、排水、上下水道铺设、土壤侵蚀管理、石块清除、杂草清除、地下公共设施以及其他任务。一旦选中了位置，保护水质、控制土壤侵蚀和沉积物都是非常重要的工作。由于许多地方需要承受风暴，所以工地上的所有侵蚀控制措施必须在砍倒第一棵树和铲掉第一锹土之前到位并做好检查。

雨水管理系统非常复杂，因为滞洪区本身就复杂且有边坡和平坦的坡底。清理工作就是清理工地道路，以便其他车辆容易通过。通常情况下清理工作的范围可由 GPS 推土机帮助完成。燃烧垃圾是传统的方法，但这种做法用得越来越少。此外，在大多数地区，空气污染标准阻止燃烧垃圾。

施工准备需要大量地规划和使用重型机械，你永远不知道什么时候你需要某种工具，到了第二天你又需要别的东西。所以在项目开始前的准备是至关重要的。挖掘中最常见的任务是清理场地、建造房屋、铺设管道、修复漏水问题、挖掘地基和需要团队进行的其他工作。

铺设管道需要技能和良好规划。沟挖好后必须确保仰角正确。管道规格必须与蓝图匹配。几种用于生活用水、污水、雨水的不同管道需要仔细铺设。

开始施工准备工作必须有许可证，取得了工地许可证才可以动工。在某些情况下，你（建筑单位）得提及施工准备工作会带来多大的干扰。为此，你还要拿到根据计划要开始动工的环境研究文件。当在现场挖掘时，建一个池塘需要挖掘和移除土，这也需要技能高超和经验丰富的工人。检修孔是一种用以维修的开口小洞或通入点，创建的目的是检修雨水渠、下水道、电话线、电力管道。

部分屋顶的施工准备工作中，暖通空调系统通常是关闭的，同样在喷涂底漆、喷涂聚氨酯泡沫塑料和涂料期间，它也是关闭的。系统关闭会阻止粉尘、气溶胶和/或蒸汽向内部空间扩散。一旦暖通空调系统关闭，就要用塑料布和胶带密封进风口，防止灰尘和喷雾进入通风口。喷涂工作完成后继续保持塑料布的密封状态至少几个小时，通常 24 小时或更多；如果进行刷涂料工作将需要更长的时间，时间长短取决于涂料是否硬化或凝固，是否散发蒸汽。直到适当的时间后塑料布和胶带已移除，暖通空调系统才重新启动。

在筹划存储保护高压安装时有很多因素要考虑。是在有人的建筑物里还是在建设中的大楼里做？大楼是否要腾空？作业过程中其他行业的工人是否在场？是在室内还是室外作业？工作区域有多大，一个开阔的地方，还是一个阁楼或通风有限的狭小空间？

总之，开始在工地工作时，你需要考虑工地本身的情况、规则和条例、急救措施和团队状况等方面。这有可能是件很有趣的工作，条件是你可以很容易地处理重型机械并且具有团队精神。

Passage 2：施工规划过程的三个阶段

Though there are numerous possible plans available for any given project, forming a good construction plan is exceptionally challenging. While past experience is a good guide to construction planning, each project is likely to have its special problems or opportunities that may require considerable ingenuity and creativity. Furthermore, the construction process is dynamic, which means the construction plan has to be modified as the construction proceeds.

Normally, the construction planning process consists of three stages: the estimating stage, the monitoring and controlling stage and the evaluating stage. It begins when a planner starts planning for the construction of a project and ends when the evaluation of the outcome of the construction process is finished.

The estimating stage involves implementing costs and duration estimate for the construction. In this stage, resource requirements for the necessary activities will also be estimated by a planner. A careful and thorough analysis of different conditions brought by site characteristics will be carried out to determine the best estimate.

In the monitoring and controlling stage of the construction planning process, the construction manager has to keep constant track of both durations and costs of the ongoing activities. Sometimes the construction is on schedule or ahead of schedule, but costs are not within the estimate or below the estimate. Therefore, constant supervision is necessary until the construction is completed. When the construction process is finished and information about it is provided to the planner, the third stage of the construction planning process can begin.

The evaluating stage is the one in which the results of the construction process are evaluated against the estimate. A planner is sure to be aware of and handle the uncertainties during the estimating stage. Only when the outcome of the construction process is known, can he/she evaluate the validity of the estimate. Also, it is in the last stage of the construction planning process that he/she can determine whether the estimate is correct or not. If it is not, he/she should make adjustments in future construction planning.

第六节　课后练习

I. Technical Terms Interpretation

1. construction streamline method

2. floor space under construction

3. program budgeting

4. project supporting services

5. construction plan

6. construction period

7. construction enterprises

8. preconstruction stage

9. construction area

10. constructional deficiency

11. 施工图预算

12. 施工文件

13. 施工误差

14. 施工现场

15. 施工详图

16. 项目筹备融通资金

17. 项目贷款

18. 项目单

19. 项目的拟定

20. 项目的总投标价值

Ⅱ. Sentences Interpretation

1. We should work according to the overall schedule chart (the construction time schedule) of the project.

2. The effective date of this contract will begin from Dec. 30th, 2027.

3. The Seller will provide preliminary (final) technical documents for Buyer in May.

4. Our major planning items contain estimating of cost and construction schedule.

5. We shall have a design collecting (preliminary design, final design) meeting next month.

6. Field erection work (civil work) will begin in October this year and complete on June first next year.

7. The date of acceptance of this plant will be April sixth, 2028.

8. The Seller's operating group (A crew of specialists) will remain on the job until guarantees are met.

9. We must take the plant through the test run and finally into commercial operation.

10. We have to change our plan for lack of materials (construction machinery, erection tools).

Ⅲ. Dialogue Interpretation

张天:Hi, Xu Hui. How are you these days? Sorry, I didn't have much time to talk to you.

Xu Hui:我很好。你在忙什么?

张天:I'm busy looking for projects recently.

Xu Hui:有进展吗? 你知道,我是新来的,我想尽快进入一个项目。

张天:Nothing much in progress. But this is quite normal.

Xu Hui:哦,我原以为这里会有很多项目,因为利比里亚是一个发展中国家,而且发展速度很快。

张天:You're right, so we have many chances. Before that, I think you can try to gain more project-related knowledge first.

Xu Hui:这就是我在想的,我只是想问你,在一个项目开始之前,我们应该做什么。

张天:Well, any project has a pre-construction stage which consists of planning, budgeting and obtaining permission.

Xu Hui:是的,我对这些有所了解。规划是一个如此复杂的过程,通常需要数年时间,尤其是当涉及公共资金时。

张天:Yes. It seems that you know much about a project.

Xu Hui:不,我是一个纸上谈兵的人,我从来没有真正接触过任何项目。因此,还需要更多的现场工作。

张天:Stop being too humble. I'm free this afternoon. Why don't you join me for a cup of coffee in Royal Hotel?

Xu Hui:我愿意,我们可以继续这个项目的话题吗?

张天:Certainly.

Ⅳ. Passage Interpretation

Main Parts of Pre-construction Preparation

Pre-construction preparation usually includes technical preparation, material preparation, labor preparation, construction site preparation and off-site preparation. Each will be illustrated in detail as follows.

Technical Preparation:

It is the core of the pre-construction preparation. Any technical error may cause serious accidents, resulting in huge loss of life, property and economy. Therefore, technical preparation must be done meticulously. Generally speaking, technical preparation includes reviewing and familiarizing yourself with the construction draw-ings and relevant design data; investigating and analysing the original data of the construction site; establishing data management system, measurement and control system and quality inspection system, and so on.

Material Preparation:

Materials are the material bases for the smooth construction of the project. The preparation of materials, such as building materials, construction and installation equipment, production and processing equipment, etc., as well as the processing preparation of components must be completed before the project starts. When preparing the materials, you should confirm the sources, properly plan the transportation and arrange the storage according to different requirements of different materials so that to meet the construction needs.

Labour Preparation:

In terms of scope, there may be labor preparation for the whole construction enterprise, large-scale comprehensive construction projects, and small and simple unit projects. In terms of content, labor preparation includes setting up a leading organization of the proposed project; organizing a competent construction team; clarifying the related design, planning and technology of construction to the construction team; establishing and perfecting the administration system.

Construction Site Preparation:

A construction site is a space in which all participants work in a rhythmic, balanced and continuous way to achieve the goal of high quality, high speed and low consumption. The preparation of construction site is mainly to create and guarantee favorable construction and material conditions for the proposed project. It includes making good control network measurement at the construction site; constructing infrastructure like water and electricity; building roads and leveling the ground; carrying out supplementary exploration on the construction site; establishing temporary facilities; installing and testing machines and tools, and so on.

第六章　工程材料采购与设备租赁(参考答案)

第三节　对话口译

Dialogue 1

A:打扰一下,先生,请问您的店卖钢筋吗?

B:Yes. We supply rebar with various diameters and strength. Which kind of rebar do you need, round ones or deformed ones?

A：我要用 6 毫米直径的光圆钢筋和直径大一点的螺纹钢筋，你能告诉我这些钢筋的价格吗？

B：Of course, you see, here is the current retail price list. Generally speaking, the prices are ＄225 per ton for round bar and ＄230 per ton for deformed bar.

A：批发价呢？

B：The wholesale price depends on the quantity of the batch you order, the bigger the quantity you order, the cheaper the price you will be offered. How many tons do you need?

A：如果价格合适，我们的项目至少要 5000 吨。

B：Up to this quantity, I would offer you ＄210 per ton for round bar and ＄215 per ton for deformed bar.

A：我想报价已经包括到现场的运输费了。

B：It's difficult to promise you. Where is your site?

A：离这 15 km。

B：In this case, you have to bear the transportation fee yourself.

A：我在哪儿能买到碎石和沙子？

B：You can buy them from the quarry at the bottom of the hill.

A：谢谢。

Dialogue 2

H：Mr. Liu, someone told me that you are responsible for the equipment leasing business of this company. I'm here to see if we can rent some of the equipment we need from you.

L：你真是来对地方了。我们有大量进口的重型施工设备，可以随时租给你。

H：What equipment do you have?

L：我们有推土机、压路机、挖掘机、电动平地机、自卸卡车和装载机，等等。

H：I am thinking of hiring an excavator and two road rollers this time. What is the volume of your excavator's bucket?

L：挖掘机卡特乐尔 235 型铲斗的容量为 2.8 立方米。

H：What are their hire rates respectively?

L：这看你是以什么方式租的，是想要按月租，按小时租，还是按平方米租。如果按月租用，挖掘机每月 3 万元，压路机每月 2.4 万元；按小时计费，挖掘机每小时 30 元,压路机每小时 25 元;按平方米收费,挖掘机每平方米 5 元,压路机每平方米 3 元。

H:Does the rate include fuel and the operator?

L:是的,还包括设备维修费。

H:Who will be responsible for the transportation of the equipment from your depot to our job site?

L:如果租用期限超过 280 小时或一个月,我们将自费用拖车运设备。否则,每运一台机器,我们加收 500 元的运费。

H:How do you usually record working hours?

L:我们的操作员填写计时单,再由你们的现场管理人员会签。计时单我们有固定格式,每份计时单一式两份,双方各持一份。计时单为支付的依据。

H:Talking about payments, how often do you want them to be made? Which payment method do you want?

L:每周末。我们需要你们先付一些预付款,最好通过电汇的方式付款。

H:We prefer to pay you monthly, because we receive payment from the project owner every month. We plan to use your equipment for three months this time. It's not a short period.

L:可以。你们的意思是你们将来会租更多的设备吗?

H:Yes, and it is likely to be a long-term lease, if your current rental fee is reduced by 18%. Due to changes in construction design, our excavation workload has greatly increased. We are considering whether to purchase or rent the additional equipment needed.

L:为了表示我们的友谊以及与你们建立良好工作关系的愿望,我愿意在报价的基础上给你们 4% 的折扣。如果你们从我们这里租赁设备超过三个月,我们将降低 15% 的租费。

H:Great! Could you please deliver the equipment to our site next Monday?

L:没问题!

第四节　篇章口译

Passage 1:材料采购和租赁会议

大家好,欢迎参加今天的会议,我们将讨论有关材料采购和租赁的重要事项。

首先,让我们来谈一谈材料采购。及时采购高质量的材料对项目的成功至关重要。我们需要确保从可靠的供应商那里获得所需的原材料,并在计划的时间内交付。为了降低成本并提高效率,我们计划与几家优质供应商建立战略合作关系。

其次,关于租赁方面的问题。有时候,我们可能需要临时租赁设备或场地,以满足项目的需要。在选择租赁方案时,我们将重点考虑设备的适用性、租赁期限和费用。我们希望通过租赁服务,灵活应对项目的变化需求。

在做出决策之前,我们将进行详尽的市场调研,以确保我们选择的供应商和租赁方案符合项目的要求,并能够提供卓越的服务。

谢谢大家的关注,让我们共同努力确保项目的成功!

Passage 2:采购步骤

Matters that need to be standardized in the procurement workflow:

1. All procurement requests must be filled out in quadruplicate. After being approved and signed by the department manager, the entire quadruplicate must be submitted to the asset accountant for review before being sent to the director.

2. The purchase request form consists of four copies, and after approval, the first copy will be used for warehouse receipt. The second copy is archived by the procurement department and organized for procurement. The third copy shall be archived and verified by the cost accountant of the finance department. The fourth copy is archived by the department.

3. Review the procurement application list: After receiving the procurement application form, the procurement department should conduct the following review to prevent errors and omissions.

(1) Sign off verification. Check whether the purchase request form has been signed off by the department manager and verify its correctness.

(2) Verify the quantity. Review the inventory quantity and monthly consumption, and determine whether the quantity on the purchase request form is correct.

4. Invite suppliers to provide quotations.

Passage 3:周转材料须知

周转材料租赁管理

1.公司周转材料实行两级管理,一级核算。周转材料租赁部是公司周转材料内部租赁市场的供方,负责周转材料的采购、租赁、收发、保管、维修、核算等工作。公司所属项目部不得私自购买周转材料,项目工程所需周转材料必须从周转材料租赁部租用,不得直接从外面租,由周转材料租赁部统一进行内外调节,满足项目施工需要。

2.进行集中管理的周转材料包括:

(1)保温箔片;

(2)隔热保温材料;

(3)窗铁栅、窗插销、窗开关调节器;

(4)乙烯基树脂涂料涂层;

(5)屋面材料、防潮材料等。

3.各项目部对新开工程应按施工组织设计(或施工方案)编制单位工程一次性备料计划。周转材料进场前,提前半个月向租赁部申报,并根据《合同法》及施工管理有关规定与周转材料租赁部签订租赁合同,明确双方责任、权利、业务和租费、维修费、赔偿费、运费等收费标准及材料进出的验收方式。租赁部应按合同规定提前做好供应准备。

4.项目部应根据施工进度,按月编制周转材料每周进出场计划(主要内容包括使用时间、数量、配套规格等),并经项目负责人签字后报送租赁部。各种计划与实际用量应基本相符,计划与实际用量不相符,所造成的经济损失,由项目部全部承担。

5.项目部和周转材料租赁部均应建立周转材料台账,做到名称、规格、进出场日期、数量准确,最后验收和结算。

第六节　课后练习

Ⅰ. Technical Terms Interpretation

1. procurement plan

2. deferred payment

3. accept order

4. after service

5. alternative materials

6. alternative bid

7. harbor and port dues

8. remittance

9. bank draft

10. liquidated damages

11. 预估每年需求量

12. 缺陷责任期

13. 预付款

14. 投标保证金

15. 到期债款

16. 履约保函

17. 出口许可证申请书

18. 连署

19. 赔偿保证书

20. 滞期费

Ⅱ. Sentences Interpretation

1. Bolt (screw, nut, stud, spring washer, pin, ball bearing, roller bearing) is the most commonly used machine part.

2. There are four broad classifications of steel: carbon steels, alloy steels, high-strength low-alloy steels and stainless steels.

3. Typical structural steel shapes include beams, channels, angles and tees.

4. The first step in procurement is inquiry, including in the contractor's home country, the country where the project is located, and a third country.

5. Building materials should be stored in specialized material warehouses that meet the requirements based on their different properties. They should be kept away from moisture, rain, explosion, and corrosion.

Ⅲ. Dialogue Interpretation

A: Hi, Mr. Brown, we have visited the factories and I'm very satisfied with your factory's production conditions.

B: 是的,这些厂生产经验丰富,历史悠久,是我们出口工具的主要生产厂家。他们所有的产品在国际市场上都享有很高的声誉。

A: That's very good. But I have little knowledge about the packing of your pliers. I just want to know the details about that.

B: 可以,我带你去看看包装。在一楼,我们有一间陈列室。我们一起下去看一下吧。

A: Sure.

(Now they are in the showroom. They are walking around the room looking at the samples of packaging. Mr. Brown is explaining the packing to the customer.)

B:这些是钳子的各种包装。通常有三种:薄膜包装、挂式包装、罩板包装。

A:Oh, the packing looks very nice.

B:薄膜包装是这种产品在国际市场上的最新包装。它惹人注目,能帮助促销。

A:Good, what about the export packing?

B:每两打装一盒,一百盒装一木箱。

Ⅳ. Passage Interpretation

货物采购是指承包商通过招标、询价等形式选择合格的供应商,购买工程项目所需要的材料、设备等货物。在实施国际工程货物采购时,还需了解国际市场的货物价格、供求关系、货源、外汇市场、保险和运输及国际贸易情况等相关信息。货物采购管理包括从项目外部采购或获取所需材料、设备、服务或其他成果的各个过程。这些过程包括编制和执行货物采购协议所需的管理和控制过程,如采购合同、订购单、协议备忘录和服务水平协议等。

第七章　工程项目管理之进度管理(参考答案)

第三节　对话口译

Dialogue 1:审议和商谈大坝工程的进度

R:Good morning, everyone. Welcome to today's meeting. We will review and discuss the progress of the dam construction project. Mr. Liu, please start by giving us updates.

L:早上好,勒努先生。我很高兴向您汇报大坝工程的最新进展。目前,我们已经完成了地质勘探和设计阶段,并顺利进入了土建施工阶段。本月初,我们已经完成了地基处理工作,并已开始进行钢筋混凝土结构的施工。

R:Excellent, Mr. Liu. Is the civil construction phase proceeding as per the plan? Are there any key milestones that we need to pay special attention to?

L:是的,土建施工目前正按计划进行。我们已经制定了详细的施工计划,并按照施工图纸和工程规范进行作业。关于关键节点,接下来我们需要特别关注钢筋混凝土浇筑和固化、大型设备的安装以及土石方工程的进展。

R:Alright, these are all crucial stages of construction. How is the progress of pouring and curing?

L:浇筑和固化是比较复杂的工作,需要充分考虑混凝土的质量和固化时间。我们将尽力控制施工进度,确保每个浇筑节点的质量和时间。目前,我们已经完成了第一次浇筑,接下来将进行后续的浇筑工作。

W:In addition to the schedule, we also need to pay attention to quality control and safety management during the construction process. We have developed detailed quality control plans and safety measures to ensure the quality of the project and the safety of the construction site.

R:这些都非常重要。我希望大家继续保持高度的责任心和专业精神,在施工过程中紧密合作,确保工程的顺利进行。刘先生,你们还有其他需要商讨的事项吗?

L:At present, there are no other specific matters. We will continue to execute the work in accordance with plan and provide timely progress reports. If there are any issues or adjustments, we will communicate with you in time.

R:非常感谢刘先生和王先生。我对大坝工程的进展感到满意。我们会继续密切关注进度和质量控制。如果有任何需要讨论的事项,随时与我们联系。

L:Mr. Reno, thank you very much for your support. We will work closely together and ensure that the dam project is completed on time and with high quality.

R:好,此次会议到此结束,最后祝大家工作顺利、合作愉快。

Dialogue 2:提交月度报表

B:Today is 28th. It's the date for our submission of the monthly statement.

A:是的。如果你带来了,请提交。

B:Here it is. It shows all the quantities of work done up to the 25th of this month. They have been checked in accordance with the actual progress on site and the drawings.

A:(仔细地看)好的。这些单价出自什么地方?

B:They come from the Bill of Quantities in our tender document.

A:哦,你们的报表总体上来讲是好的。但第一项空压机的混凝土基础应该从清单中删除,因为经过我们两天前的检查,质量不合格,混凝土表面有几条裂缝。你们应该立即自费清除并重建。

B:OK.

A:这一项指的是什么?

B:This is the sum for materials and equipment delivered to site this month.

A:但根据合同,月进度报表中的到场材料量应该是进场材料和设备价格的70%。

B: Yes, we did leave 30% of the value of the materials and equipment out of the Monthly Statement, complying the Contract. And please certify the quantity of the works done under Variation Order this month.

A: 我们现在就可签证,但某些变更项的价格水平很高。你们的定价依据是什么?

B: The prices are based on the quotation from our material suppliers, because the materials are not the same as those in the original contract and their unit prices can't be fixed in accordance with the Bill of Quantities.

A: 这部分的报价需要进一步协商。我们把这个问题留在下一次讨论吧。

B: Fine. Do you have any other comments?

A: 我想提醒你的是,你的月进度报表的金额必须扣除 10% 的质量保证金。

B: Thank you for your reminder, I will give you our modified Monthly Statement on tomorrow morning.

第四节　篇章口译

Passage 1:工期滞后原因分析

大家早上好。整个工程从 2 月 10 日开工至今已有 105 个工作日,根据施工合同要求,工程应于 5 月 10 日竣工,到目前工程已超过 15 天。根据前期施工情况分析,造成工期滞后的原因大致有以下几点:

首先,施工单位总包对作业人员交底不明确,对专业施工队掌控不力,没有相应的制约措施,交叉作业安排不到位,造成部分部位重复施工。3 号楼隔墙经单层石膏板密封后,隔音材料未及时安装。在例会上曾多次要求施工单位尽快完成安装,但执行效果不太理想。目前,2 号楼吊顶内铺设有消防、弱电管路,吊顶龙骨架也已安装到位,但用于送风排风的全热换风机才刚到场。以上这些情况对工程进度造成了比较大的影响。

其次,各个楼层风管已基本安装敷设完成。各楼层的空调内机及空调冷媒管气密试验于 5 月 5 日已全部完成,目前 2 号楼、3 号楼空调外机均未到场,外机到场需吊装就位且外机的系统连通、真空试验以及调试也需要一定的时间,这也是影响进度的一个重要因素。

最后,消防安全施工审批问题也是造成 2 号楼、3 号楼进度受到影响的一个原因。本装修工程 3 月 24 日才收到消防设备的证明文件,此前为不影响工程进度,在 3 月 18 日第六次工程例会上,业主和物业同意拆除消防喷淋管。

Passage 2：实现进度目标的举措

Good morning, everyone, I am project manager Wang Peng. Next, I will give a speech on measures to achieve the progress plan goals in the future.

Firstly, we undertake to adopt the most advanced international BIM standards when undertaking your company's international major engineering projects. Throughout the entire life cycle of the project, we will utilize over 30 BIM software to optimize the entire project management process, providing assurance for project schedule control.

Secondly, we will prepare a project schedule plan based on factors such as the project implementation plan's schedule goals, resource allocation capabilities, provision of design drawings, and external conditions. Then, the specific project schedule will be further decomposed into weekly and daily plans, in order to achieve the schedule goals of weekly and monthly guarantees and annual guarantees.

Then, during the construction process, we will implement flexible and dynamic control. In the process of adjusting construction projects, the owner usually requires the contractor to submit a 2-week rolling plan, weekly construction progress report, and other documents to prove and ensure that the project progress meets the contract requirements. In this regard, we guarantee that we will promptly compare the benchmark plan of the project with the tracking plan during the process. Once any deviation is found, we will analyze the cause of the problem in a timely manner and adjust the sub item plan through various means according to the actual situation on site, so that the actual construction period matches the benchmark construction period plan.

Finally, we will strictly control safety and quality. We will always adhere to the principle of "safety first, quality first", and strengthen the supervision, inspection, and guidance of project safety and quality through optimizing institutional processes and increasing project evaluation in project process management, in order to avoid the occurrence of idle work and rework. Thank you!

第六节　课后练习

Ⅰ. Technical Terms Interpretation

1. force majeure

2. double shift

3. schedule of earthworks

4. commencement of work

5. contingencies

6. hurry

7. Crucial Path Method

8. information distribution

9. quality assurance

10. responsibility assignment matrix

11. 项目相关者

12. 预期利润

13. 进度偏差

14. 次关键工作

15. 完成百分比

16. 预算费用

17. 基准计划

18. 术语

19. 周进度会

20. 竣工报告

Ⅱ. Sentences Interpretation

1. We should work according to the overall schedule chart (the construction time schedule) of the project.

2. The Seller will provide preliminary (final) technical documents for Buyer in May.

3. On our most projects, Critical Path Method (CPM) is used for scheduling.

4. The Seller's operating group (A crew of specialists) will remain on the job until guarantees are met.

5. Every month we shall establish construction schedule.

6. The date of acceptance of this plant will be April 6th, 2025.

7. This contract plant will be put in commissioning on Jan. 1st, 2028.

8. Our major planing items include cost estimation and construction schedule.

Ⅲ. Dialogue Interpretation

A: I'm sorry to say that I am not very satisfied with your construction progress. According to the total schedule of the plant renovation project, you should have completed the roof installation project by now.

L:由于天气不好,进度延迟了。你看,施工日记显示有 5 天下大雨,2 天有大风。那时候我们不得不停止安装屋顶。

A:The erection of the main processing machines must start on the 10th of next month. It's one of the activities in the critical path. If you fail to meet the key date, it will seriously affect completion and take-over. So you have to reach it. There's no room for discussion of the key dates fixed in the contract.

L:我明白。

A:What's more, your HVAC installation work seems to behind schedule.

L:工程进度落后是因为你们耽误了设备采购的审批。

A:I can't agree with you there. You want to use different equipment instead of that describe in the specifications. However, we didn't receive your written proposal until 15th of last month. After we investigated your purposed equipment carefully, we issued our confirmation to you on 26th of last month. Our response to you was still within the 2-week limitation stipulated in the contract. Here are the records of document transmittal. I suggest we discuss this later. Now, the important thing is to finish the work in the shortest possible time.

L:我们可以在五十天内完成。

A:No, that's too late. Can you do it within 45 days by working two shifts?

L:我们会尽全力的。我们将调用更多的资源,重新安排工程进度。我向您保证在 3 天之内将修改后的计划提交给您。还有一件事我要提醒您,我们想申请延长厂房附属区土方工程的时间,并且增加费用。你们修改后的平面图本应在 1 个星期前发下来,但我们现在仍未收到。如果不增加劳动力的话,即便是我们今天拿到图纸,也很难赶上正常的进度。

A:There are still some issues undetermined on the layout. We shall speed up. But it is impossible to extend the key dates, for each key date is correlated with the next one. We agree to take the additional cost into consideration. But firstly, you should submit your written application.

L:好的,我们会准备的。

Ⅳ. Passage Interpretation

让我们从引水结构开始。对于围堰,我们已经浇筑了 2600 立方米的混凝土。它将在未来几天内如期完成。在左桥台和坝基处,我们已经完成了 5000

立方米的岩石开挖。这是上一个月在引水结构上取得的成就。现在让我们看看隧道挖掘。在 A 洞,我们在上游工作面前进 120 米,在下游前进 80 米。我们还做了 590 块钢锚栓,重达 4.5 吨。在 B 洞,隧道前进 40 米。岩石螺栓,750块。B 洞口的入口加起来有 15 吨。这是掘进用的。现在来看看发电厂。地基挖掘正在那里进行,上一个月完成量为 5000 立方米。最后,碎石厂的安装已经完成,目前正在试运行。

第八章　工程项目管理之质量管理(参考答案)

第三节　对话口译

Dialogue 1:施工现场安全管理

A:我想先和这位焊工谈谈。你们有没有参加过安全教育的培训课程?

W:Yes. In class, I learned all the safety regulations related to welding.

A:戴头盔和护目镜没问题,但你的劳保鞋在哪里? 你的经理不是给了你一双吗?

W:Yes. Sorry, I forgot to wear it today.

A:不要再忘记了。你在这里获得焊接的"工作许可证"了吗?

W:Absolutely, here.

A:(通过开口)从这里我可以看到,地板上既没有护栏,也没有足够的盖子来封住这么大的开口。它不符合安全规定,应立即纠正。

W:Foreman, you have to arrange for two workers to seal the hole at once.

A:另外,请记住在此处设置警告通知,以提醒人们注意该区域的危险。

W:We will deal with it immediately in accordance with your instructions.

A:好了,今天的安全巡查就到这里了。我希望所有不符合项都能尽快得到纠正,勿再重蹈覆辙。我会在下一次安全巡查中逐一检查。谢谢。

Dialogue 2:工程细节核查

A:你完成了这个泵的安装吗?

B:Yes, it's already done.

A:情况如何?

B:Very good! The anchor bolts are vertical, the bolts are tightened, and the torque is consistent. The contact between the nut and the washer, and the washer and the base of the equipment is very tight.

A:垫铁怎么样?

B:The pad-iron group is stable and in close contact. No more than three pieces per group.

A:请说明联轴器对接端部空间的要求是多少毫米?

B:The minimum clearance can reach 20 mm.

A:这个设备安装的标高是错误的,不是吗?

B:No, the elevation meets the requirements of the drawing specification.

A:请问,这台压缩机是否需要拆解进行检查?

B:Yes.

A:我可以启动机器吗?

B:No, check it out as well.

Dialogue 3:检查安装任务

A:Good morning, everyone.

B:早上好。

A:Today, we are tasked with installing this distillation column, please prepare.

B:是的。请告诉我们这座塔的重量、长度、直径和重心。

A:The weight of this tower is 34 tons, length 43 m, diameter 9 m; The center of gravity is 15 meters from the bottom of the tower.

B:谁是安装工作的负责人?

A:Engineer Wang is in charge of the installation work today, please thoroughly inspect the tools, such as electric winch, wire rope, pulley, and figure out if they are all in good condition. Check the foundation screws (anchor bolts) against the drawings.

B:好的,我们已经检查了它们。

A:Okay, let's get to work!

W:大家注意! 听我口哨指挥。

A:All right! The tower is in place, check its verticality.

W:塔的垂直公差小于其高度的千分之一。

B:可以接受!

A:Okay, please tighten the base nut. Let's rest.

第四节　篇章口译

Passage 1

女士们,先生们,今天真是美好的一天。让我先介绍一下自己。我叫李飞。

今天,我很荣幸能与大家谈谈工程质量管理的关键方面,强调其在该领域的重要性和必要性。

工程质量管理是任何成功项目都不可或缺的组成部分。它包括一套系统的过程和实践,旨在确保项目符合规定的质量标准和要求。该方法在各行各业工程项目的整体成功和长远发展中起着举足轻重的作用。

工程质量管理至关重要的主要原因之一在于,它能够确保符合既定标准和法规。在当今全球化的世界中,项目通常涉及不同利益相关者之间的协作,每个利益相关者都有自己的一套指导方针。有效的质量管理可以保证项目符合这些标准,促进国际合作,并最大限度地降低法律纠纷的风险。

质量管理是一种强有力的风险缓解策略。识别潜在风险并采取积极措施来应对这些风险是这一过程不可或缺的部分。

通过系统地管理风险,工程项目可以避免代价高昂的错误、延误和声誉损害。这种积极主动的方法增强了项目在面对不可预见的挑战时的整体应变能力。

从长远来看,在质量管理方面进行前期投资可以节省大量成本。通过在项目的早期阶段发现并纠正质量问题,企业可以避免以后花高价去返工和修改。这不仅节省了财政资源,还确保了项目符合其预算和时间表。

工程质量管理直接有助于提高客户满意度,达到甚至超过顾客对质量期望可提高客户的满意度。

Passage 2

Good morning! I'm Wang Jian, I work in Company A, and I'm the representative of the engineering project quality management. First of all, thank you for providing this opportunity to our company. Next, I would like to tell you about our business.

Our practice covers all aspects of engineering quality management, including but not limited to ensuring that the project meets relevant standards and specifications at every stage through rigorous inspection and monitoring procedures; Provide comprehensive quality assurance services to ensure that the overall quality of the project meets the client's expectations and exceeds expectations. Adopt advanced technology and innovative methods, combined with industry best practices, to ensure that our customers receive the best engineering quality management services.

We are committed to continuous improvement and innovation to adapt to changing industry needs and standards, and our company was responsible for the quality management of a hydropower project in Jiangxi province the year before, and the project was successfully delivered, which was praised by customers, and there have been no safety accidents or power generation failures in more than two years of operation. This not only demonstrates our expertise, but also demonstrates our commitment to quality and customer satisfaction.

Overall, our company is known for its professional, reliable and innovative quality management services, and is committed to helping our customers ensure that their engineering projects meet high quality standards and are successful. We will attach some brochures with detailed information about our company after the RFP. In these brochures, you can learn more about our headquarters, as well as the quality management cases of several projects that our company has undertaken in the past.

第六节　课后练习

Ⅰ. Technical Terms Interpretation

1. 现场控制

2. 屈服点

3. 在制品

4. 抗张强度

5. 现场检查工作

6. 延伸率

7. 制程品质保证

8. 布氏硬度

9. 弹性制造系统

10. 现场制作

11. manufacturing verification test

12. profiled channel

13. total productive maintenance

14. sump pan

15. defects analysis system

16. safety officer

17. manufacturing project management

18. applied trim area

19. key performance indicate

20. material review board

Ⅱ. Sentences Interpretation

1. It is useless to only know some quality management methods without knowing how to practice them.

2. Supervision Dept shall manage construction safety, quality, schedule and cost, etc. according to contract; if there is no Supervision Dept, MSC-PE is responsible for project management.

3. After pile foundation engineering is over, the supervision company shall organize pile foundation engineering acceptance according to the related laws and regulations and obtain the document for passing the acceptance of pile foundation engineering.

4. Quality control has to be imposed by the contractor whereas quality assurance is carried out by a separate third party agency engaged by the owner.

5. The properties of concrete in the outer zone are greatly influenced by curing as it is the concrete in this zone that is subject to moisture, carbonation and abrasion.

Ⅲ. Dialogue Interpretation

A：What's written on the boom side?

B："危险！不要在臂下停留！"

A：But, I have to rope the machine.

B：你一固定好,就及时离开。

A：But, I must guide the anchor bolts in.

B：那你只好冒点险了。

A：I think the provision is useless.

B：不对。它提醒你随时注意安全。

A：Yes, I agree. "Safety first."

B：你看见过东西从高处掉下来吗?

A：No, never.

B：那是常有的事。

A：Really?

B：当然！

A：Thank you! I will keep that in mind.

Ⅳ. Passage Interpretation

影响工程项目质量的因素有很多，归纳起来主要有六个方面，即人（man）、材料（material）、机械（machine）、方法（method）、测量（measurement）、环境（environment），通常称为 5MIE 因素。对这六个因素严格控制并进行全面协调管理，是保证工程质量的关键。

人，指参与工程项目建设的决策者、管理者和操作者，其素质和能力的高低，都会影响到工程项目的质量水平。项目管理者进行质量管理时，应从实施者的素质、理论及技术水平、生理状况、心理行为等方面对人的因素加以考虑并控制。

机械是工程实施机械化的重要物质基础，项目管理者应综合考虑工程项目施工的条件和特点，制定机械化施工方案，使机械设备和施工组织方案有机联系起来，有效提高工程项目的综合效益。

方法是保证工程质量稳定提高的重要因素。项目管理者应根据工程实际，全面分析和考虑工艺、技术、操作、组织、程序等方面，力求工艺技术先进、操作正确、组织合理、程序完善，以提高工程质量、加快工程进度和降低工程成本。

测量，指工程项目施工中采取的测量方法是否标准、正确。项目管理者应注意针对测量任务所需的准确度，准备相应的测量工具，并定期检验和校准测量工具和试验设备，制定必要的校准规程，以减小测量过程中产生的误差。

环境，包括工程技术环境、工程作业环境、工程管理环境和周边环境等很多环境因素。这些环境因素复杂多变，项目管理者应加强环境管理，根据工程项目的条件和特点，采取必要的措施控制环境对质量的影响。

第九章　工程项目管理之成本管理（参考答案）

第三节　对话口译

Dialogue 1：成本管理说明

李玲：Hey, Tom, do you know anything about cost management?

Tom：是的，当然。成本管理是识别、分析和控制成本以提高企业绩效的实践。

李玲:That's interesting. How do companies carry out cost management?

Tom:公司有不同的策略来控制成本。例如,一家公司可能专注于降低其固定成本,如租金和工资,或者它可能试图最小化其可变成本,如材料和供应。

李玲:I see. What about cost analysis? Can you explain it to me?

Tom:当然。成本分析包括审查与特定项目或过程相关的所有成本,这样你就可以确定哪些成本是必要的,哪些是不必要的。一旦你做到了这一点,你就可以开始探索尽可能降低成本的方法。

李玲:That makes sense. Does the company have tools to manage costs?

Tom:是的,有很多不同的工具和技术。例如,许多公司使用成本会计软件来跟踪他们的开支并生成报告。他们还可能使用成本效益分析来确定某项投资是否值得。

李玲:All right. How does cost management affect the company's profit?

Tom:嗯,有效的成本管理可以帮助公司提高盈利能力,因为它可以让公司减少开支,增加收入。它还可以帮助公司保持竞争力,因为它允许公司以比竞争对手更低的价格提供产品或服务。

李玲:OK, Tom, thank you for explaining it to me.

Tom:不用谢,李玲。我总是乐意帮忙。

Dialogue 2:项目成本管理会议

王林:Hello everyone, thank you for attending this project cost management meeting. We need to ensure that our project can proceed as planned and within budget.

Jerry:是的,这个项目的技术要求很高,我们需要保证在不牺牲质量的前提下控制成本。

Mary:我已经准备好了预算表,我们可以从每个阶段开始讨论。

王林:Okay, let's start from the beginning of the project. Jerry, what do you think may increase costs in the early stages of the project?

杰里:主要是研发阶段的实验和测试费用,我们可能需要购买一些特殊设备。

Mary:我会把这些放进预算表。此外,还有其他潜在成本吗?

Jerry:还有培训费用,因为我们可能需要团队成员掌握一些新技术。

王林:Okay, these are all very important. Now let's take a look at Jerry during the project execution phase. What is your opinion on the cost of the execution phase?

Jerry:首先是劳动力成本,包括工程师和技术人员的工资。此外,材料和设备的采购成本也是一个考虑因素。

Mary:我会把这些详细列在表格里。还有其他因素会影响实施阶段的成本吗?

Jerry:可能会有一些未知的技术挑战,导致额外的研发和测试成本。我们需要在预算中留有一定的灵活度来处理这些情况。

王林:I understand. Finally, let's take a look at the project completion phase. What costs may emerge at this stage?

Jerry:可能会有一些额外的调试和优化工作,以确保项目的最终交付符合客户的期望。

Mary:好的,还有什么需要考虑的吗?

王林:For example, the support and maintenance costs after project delivery. We need to ensure that customers receive timely support when using our products.

Jerry:是的,这也是一个重要的方面。我们需要在预算中为售后服务预留一些资源。

王林:Thank you very much for your feedback. Mary, could you please organize all the costs we discussed in the table?

Mary:当然,我会在表格中反映所有的成本预测,并确保我们有足够的预算来满足各个阶段的需求。

王林:Very good, thank you all for your participation. We will continue to monitor costs and make adjustments as needed to ensure the successful completion of the project.

第四节　篇章口译

Passage 1:成本管理工作总结

尊敬的各位领导、亲爱的团队成员们:

大家好! 很高兴在这个特别的时刻,与大家一同分享我们在海外基础设施建设项目中的成本管理情况。今天是我们项目启动的一周年纪念日。回顾过去的一年,我们不仅取得了许多进展,也面临了一些挑战,但通过团队的共同努力,我们已经取得了显著的成绩。

首先,我想强调成本管理在我们项目中的重要性。在国际市场竞争日益激烈的环境下,我们必须更加注重成本效益,确保我们的项目既能够高效完成,又能够保持竞争力。成本管理不仅仅是一个财务问题,更是项目的生命线,关系到我们公司在国际市场上的声誉和竞争优势。

在过去的一年里,我们一直在不断努力提高成本管理水平。我们采取了一系列措施,通过这些措施,我们已经成功降低了一些不必要的开支,并提高了项目的整体效益。

第一,我要强调我们对成本的严格管理。我们清楚地认识到成本控制不仅仅关乎财务,更是确保项目的可持续性和成功的重要因素。我们致力于在不影响项目质量和进度的前提下,精准掌控每一笔支出。这意味着我们不仅仅要关注当前的支出,更要注重长期投资的回报和效益。

第二,我们采取了一系列创新的成本管理措施。从项目启动之初,我们便建立了严格的成本核算和监控机制,确保资金使用的透明度和高效性。我们利用先进的技术和数据分析工具,及时识别潜在的成本风险,并迅速采取措施予以解决,以避免成本超支的情况发生。

第三,我们还重视团队合作与沟通。成本管理不仅仅是财务团队的责任,也是所有项目参与者的共同使命。我们鼓励团队成员沟通与协作,确保每个人都能意识到成本控制的重要性,并积极参与到成本管理的实践中来。

然而,我们也要清醒地认识到,成本管理是一个持续改进的过程,需要我们不断总结经验、发现问题并及时调整。在未来的项目实施中,我们将进一步加强与供应商的合作,深化成本核算和分析,找到更多的节约空间。同时,我们也将注重团队成员的培训,提高整个团队的成本管理意识,使每个人都能在自己的岗位上发挥积极作用。

最后,我要感谢每一位团队成员在过去一年中的辛勤付出。正是因为大家的努力,我们才能够取得今天的成绩。希望在未来的日子里,我们能够一如既往地团结协作,共同克服各种困难,为项目的成功而努力奋斗。

谢谢大家!

Passage 2:有关项目成本管理的建议

Hello everyone, I will explore project cost management based on my own work practice and the current situation of our project cost management, as well as the actual situation of our enterprise. I will propose methods and suggestions to strengthen project cost control.

The first is cost control during the bidding and signing stages.

With the development of the market economy, construction enterprises are in a tense state of "finding rice to cook", busy searching for information, busy bidding, and busy finding relationships. In order to win the bid, the construction company lowered the bid price. Some engineering projects may incur losses if management is slightly relaxed, while others may incur significant losses. Therefore, it is particularly important to do a good job in pre-bid cost prediction, scientifically and reasonably calculate bidding prices and bidding decisions. Therefore, when bidding and quoting, it is necessary to carefully identify every economic provisions involved in the bidding documents, know about the creditworthiness and contractual capacity of the owner, and have a certain degree of confidence before bidding. After completion, before quoting, relevant professionals should be organized to conduct a review and argumentation. Based on this, the final decision should be made by the enterprise leadership.

To make a good pre-bid cost prediction, enterprises should continuously collect, organize, and improve their internal price system based on market conditions, which is in line with the actual situation of the enterprise, providing strong guarantees for fast and accurate pre-bid cost prediction. Meanwhile, bidding also incurs various expenses, including bidding fees, travel expenses, consulting fees, office expenses, hospitality expenses, and so on. Therefore, improving the winning bid rate and saving bidding costs has also become an important aspect of reducing cost expenditures. The bidding cost should be linked to the indicators related to the winning bid price, implement total amount control, standardize the scope and amount of expenses, and have a dedicated enterprise leader responsible for this bidding work and management. After winning the bid, the enterprise should strive for reasonable reasons when signing the contract, especially for the current developer. The contract conditions that are not conducive to the construction enterprise have already entered the bidding documents during the bidding stage, and the construction enterprise has confirmed the bidding documents during the bidding process. It is very difficult to change them. However, we should also make full use of this opportunity to sign a contract, negotiate with the owner as much as possible on any unfavorable terms, and strive to be fair and reasonable as much as possible. We should strive to minimize risks before signing with the owner.

After signing the contract, the company should carefully organize contract disclosure to relevant personnel of the government and project departments. Through different forms of disclosure, the relevant management personnel of the project department should clarify all relevant terms and contents of this construction contract, laying a foundation for expanding the profit points of project management and reducing project losses in the next step.

The second is cost control during the construction preparation stage.

Based on the design drawings and relevant technical information, combined with the actual situation of this project, we vigorously promote the application of new technologies and processes, conduct careful research and analysis on construction methods, construction procedures, operation organization forms, mechanical equipment selection, technical organization measures, etc., and develop a scientifically advanced and economically reasonable construction plan.

The third is cost control during the construction process, which includes four aspects of content.

1. To control labor, materials, machinery, and on-site management costs well

The cost of labor is mainly controlled from the perspective of subcontracting labor costs. According to the actual situation of the project, labor construction subcontractors should be selected, and labor subcontracting prices should be determined through bidding. For auxiliary materials, small machinery, and sporadic employment on site, including garbage removal, which are not large and difficult to control, a lump sum cost can be adopted.

The control of material cost includes two aspects: material quantity control and material price control. The material usage should adhere to a quota-based material requisition system, and promote the use of new technologies, processes, and materials to reduce material consumption. We should adhere to the recycling of surplus materials, reduce material consumption levels, and reduce stacking and storage losses. When purchasing materials, purchase price control should be implemented, and a market competition mechanism should be introduced to purchase bulk materials through competitive bidding while ensuring quality and quantity. Including the use of ready-mixed concrete, the price should be determined through competitive bidding, and the quantity should be strictly controlled and settled according to the construction drawings. All other materials entering the construction site must have strict collection and storage systems to ensure quality and quantity. At the same time, it is necessary to consider the time value of funds, reduce capital occupation, and lower inventory costs. For auxiliary materials that are difficult to control and manage, after calculation, they can be included in the labor subcontracting price in a lump sum manner and used by the labor subcontractor.

Other expenses mainly include engineering water and electricity fees, garbage removal and other direct expenses. In this regard, we should adhere to the principle of conservation, implement a contracting mechanism, and implement fee management for various subcontractors on site to reduce expenses and save costs.

On site management expenses (also known as project indirect expenses) include temporary facility expenses and on-site expenses, with the main expenses being temporary facilities, project wages, transportation expenses, and business entertainment expenses. To reduce this expense, it is necessary to develop an expense plan, streamline management personnel, and implement total amount control; Strictly control entertainment expenses, and assign responsible departments and personnel to each expense according to its nature; For special expenses and large amounts of expenses, a meeting should be held to discuss and submit them to the company leadership for approval. Temporary facilities should be arranged and constructed based on the actual situation of the project and the site, with the principles of conservation, scientific rationality, practicality as much as possible, and the ability to be reused multiple times.

2. We need to strengthen construction organization, make reasonable use of resources, and reduce project costs

Under a reasonable construction period, the project cost expenditure is relatively low. The advance or delay of the construction period compared to a reasonable one means an increase in project costs. Therefore, when arranging the construction period, the project manager should consider the dialectical unity between the construction period and cost, organize balanced construction, in order to ensure the construction period and reduce costs while using resources reasonably (if the construction period is required to be advanced or delayed due to the construction party's reasons, negotiations and claims must be handled with the construction party)

3. Strengthen safety and quality management, control safety and quality costs

The cost of safety and quality includes two aspects, namely control cost and accident cost. It is necessary to implement safe production, civilized construction, improve product quality, and appropriately control costs. What needs to be reduced is the accident cost, which is the loss of the project in the event of an accident. Accident cost is the cost that disappears when there are no safety or quality accidents in the project. We must strengthen safety and quality management to minimize the accident cost of safety and quality.

4. Strengthen subcontracting management and control subcontracting costs

For subcontracting projects (labor and various professional subcontracting), in addition to strict qualification review of the subcontracting team, it is necessary to fully introduce the competitive mechanism of the market economy, implement bidding and tendering, and scientifically and reasonably determine the price of subcontracting projects. Strict management measures should be adopted, including contract signing, payment of advance payment and engineering funds, retention of letter of guarantee and quality guarantee deposits. On site, comprehensive monitoring and management should be implemented more strictly. Timely handle subcontracting settlement and lock in subcontracting costs after completion.

The fourth is cost control during the completion acceptance stage.

From a practical perspective, many projects, from the start to the completion of the final stage, transfer the main technical strength to other ongoing projects, resulting in delayed completion work and a long front line. Machinery and equipment cannot be transferred, and cost expenses continue to occur, gradually leading to the loss of economic benefits that have been achieved. Therefore, it is necessary to carefully arrange and strive to shorten the final completion time to the minimum, in order to reduce the cost expenditure during the completion stage. Special attention should be paid to the completion acceptance work. Before the acceptance, various written materials required for acceptance should be prepared and sent to Party A for reference; The opinions raised by Party A during the acceptance shall be carefully handled according to the design requirements and contract content to ensure smooth delivery.

第六节 课后练习

Ⅰ. Technical Terms Interpretation

1. delay compensation

2. project progress payment

3. proforma invoice

4. investment deviation

5. quantified claim

6. project economic dispute

7. final settlement price

8. contractor's all risks (CAR)

9. cash flow

10. quality warranty payment

11. 实现价值

12. 实际成本

13. 竣工预算

14. 竣工估算偏差

15. 成本偏差

16. 进度偏差

17. 成本绩效指数

18. 期中付款凭证

19. 劳工赔偿险

20. 无息贷款

Ⅱ. Sentences Interpretation

1. I have a look at the details. Your budget is too high. Can you reduce the cost again?

2. Reasonable selection of mechanical equipment is of great significance to cost management, especially for high-rise buildings.

3. The project department should control the use fee of construction machinery from the aspects of reducing the idle equipment caused by improper arrangement, improving the utilization rate of equipment, and avoiding the stoppage of mechanical equipment caused by improper use.

4. On-site funds should be streamlined, construction quality management should be strengthened, management level should be continuously improved, and management expenses should be reduced.

5. Technology and economy are interdependent. The improvement of technology or the adoption of new technology will inevitably greatly improve labor efficiency and save materials, thus saving costs.

6. For example, if the ceiling surface is very flat, it can meet the use requirements without plastering, and it can also increase the use space and save costs by scraping putty.

7. According to the assessment system specified by the project department, the relevant personnel responsible for cost management are assessed, and the focus of the assessment is to complete the four indicators of workload, materials, labor costs and machinery usage fees.

8. In order to effectively control the construction subcontracting fee, the project department mainly makes inquiries for subcontracting projects, concludes equal and mutually beneficial subcontracting contracts, and establishes a stable subcontracting relationship network.

9. The purchase price of materials and equipment should be controlled mainly by grasping market information and applying bidding and inquiry.

10. There are many information in the technical proposal, which including: process flow, process description, capacity of the plant, performance of the product.

Ⅲ. Dialogue Interpretation

采购经理 A:最近我们的成本超出了可控范围,供应商的价格越来越高,让我们压力越来越大。

总经理 B:This will definitely not work. We should try to control the cost.

采购经理 A:是的,我想我们可以和供应商重新谈判合同,寻求更优惠的条款。这可能包括重新谈判价格、优化交货时间、数量折扣等。

总经理 B：Good idea! However, we also have to consider other countermeasures, because negotiations may not agree.

采购经理 A：那么我们可以更换原有的供应商，寻找新的供应商，或者考虑与多个供应商建立合作关系，减少对单一供应商的依赖。这样，我们可以更好地协商价格，并在供应商出现问题时有一个替代计划可供选择。

总经理 B：Yes, I think it still depends on factors such as quality and service, not necessarily low price. We should evaluate the price, quality and service of the original suppliers, compare with other suppliers, select the supplier with the highest cost performance and cooperate with them. At the same time, the company also needs to communicate with suppliers to find preferential deductions or other ways of cooperation to reduce costs.

采购经理 A：我觉得你说得很有道理，我会和同事们一起制定一个合理的方案。

Ⅳ. Passage Interpretation

施工成本计划是根据施工项目的具体情况制定的施工成本控制方案，既包括预定的具体成本控制目标，又包括实现控制目标的措施和规划，是施工成本控制的指导性文件。项目部要在一项工程施工前认真编制施工组织设计，优选施工方案。施工方案主要包括四个方面内容：施工方法的确定、施工机具的选择、施工顺序的安排和流水施工的组织。施工方案的不同，工期就会不同、所需机具也就不同。因此施工方案的优化选择是项目施工中降低工程成本的主要途径。

第十章　工程项目验收与付款（参考答案）

第三节　对话口译

Dialogue 1：在工程验收会议上，业主和承包商代表的一次对话

A：Dear project manager, welcome to our project acceptance meeting. Please confirm whether our acceptance time and location meet your requirements.

B：非常感谢。时间和地点都很合适。让我们开始验收过程吧。

A：Fine. Let's start with the progress of the project. Please specify whether the project construction has been completed within the scheduled time.

B：是的，施工工作已按计划进行，并基本完成。但还有些扫尾工作，比如恢复地表面等。

A:My idea is that the taking-over process will be divided into two phases: First, our expert panel, accompanied by your technical staff, will inspect the completed works and see if they are satisfactory. If everything is all right, we'll enter into the second phase, which is to carry out the performance tests of the equipment and instruments as specified in the Contract.

B:我赞同。

A:Next, let's talk about the economic benefits of the project. Please explain if there were any cost overruns or savings during the implementation and provide an explanation.

B:在实施过程中,我们严格控制了成本,并且没有发生超支情况。在某些方面,我们也采取了合理的措施实现了成本节约。

A:Very good. Finally, please explain the sustainability and environmental impact of the project.

B:我们在工程设计和实施过程中充分考虑了可持续性和环境保护因素,采用了低碳、环保的技术和材料,尽量减少对环境的影响。

A:Thanks very much for your introduction. Based on our discussion, I can determine that this project meets our requirements and expectations. Thank you so much for your hard work.

B:谢谢您的认可。我们也非常感谢您对我们工作的支持。如有需要,我们将提供持续的技术支持和维护服务。

Dialogue 2:关于支付方式的对话

Y:We would prefer you to pay in US dollars.

L:我们希望贵方能够接受除美元以外的其他货币。贵方同意60日信用期吗?

Y:Sorry, we require full payment within 45 days. If you use cash payment, we offer a 10% discount.

L:但是信用证付款是我们在此类商品交易中采用的方式。

Y:We could ship your order within ten days of receiving your payment.

L:由于这批货物很昂贵,所以我方希望在交货以后付款。

Y:We require immediate payment upon freight documents.

L:由于我方有些财务上的困难,希望贵方能允许我方分期付款,在交货后先支付第一笔款项,余下的货款则每月一次支付。

Y:Sorry, we do not support installment payments. We would like to receive full payment as we also need to pay the manufacturer.

L:贵方可否为我方的这次生意破一次例,接受付款交单或承兑交单?

Y:We regret to say that we are unable to consider your request for payment under D/A terms. You can pay the deposit in cash first, and the remaining money can be settled within 45 working days.

L:我们只能接受 20% 本地货币现金支付,其余 80% 应以信用证在交货期前 15 日到 30 日开出。

第四节 篇章口译

Passage 1:对颁发完工证明的解释

尊敬的承包方代表,关于您提出颁发完工证明的要求,我们遗憾地通知您,我们无法同意此要求。我会详细解释我们的理由。

首先,根据我们在合同中的约定,完工证明书需要在工程项目正式验收合格后才能颁发。然而,目前我们还未进行工程项目的正式验收,因此无法提供完工证明。

其次,完工证明书是一项重要的文件,它代表了工程项目的正式完工和交付。作为业主方,我们需要确保工程项目的质量、安全和经济效益满足我们的要求和期望。只有在对工程项目进行全面评估和验收后,我们才能准确地评判其是否符合合同约定的要求,并在此基础上决定是否颁发完工证明。否则,过早地颁发完工证明会存在一定的风险,可能会导致在验收后发现问题或潜在缺陷,给业主方带来不必要的损失和风险。

最后,我们希望保持合同规定的程序和流程的严谨性和公正性。通过基于全面的验收评估来决定是否颁发完工证明,不仅可以确保项目的质量和安全,也能保护业主方的合法权益。

综上所述,由于目前工程项目尚未正式验收,我们无法满足您的要求颁发完工证明。我们希望您能够理解我们的立场,并愿意配合我们进行工程项目的正式验收。一旦工程项目通过验收并满足合同约定的要求,我们将立即颁发完工证明,并确保能够顺利接收该项目。

Passage 2:土木工程合同中的付款流程和须知

Payment method

1. Advance payment

In the earthwork contract, both parties can agree to pay a certain proportion of advance payment before the start of construction. This helps the construction party to enter the site in a timely manner and prepay some labor and material costs. The advance payment amount can be determined based on the project scale and mutual consultation, but generally does not exceed 30% of the total contract amount.

2. Progress payment

One of the common payment methods for earthwork engineering is to pay according to the progress of the project. When the project reaches a certain progress, the owner shall pay the corresponding amount to the construction party according to the agreed proportion. This approach is conducive to motivating the construction party to actively promote the project progress and ensure that the project is completed on time. At the same time, it also avoids the problem of capital flow for the construction party.

3. Completion payment

When the earthworks are completed, the owner needs to pay the remaining completion fee to the construction party. The completion payment is usually the final payment after deducting the advance payment and progress payment, and it is also the final compensation received by the construction party. The contract should clearly specify the payment time and method for the completion payment to ensure that the construction party receives corresponding returns.

Supplementary information on payment methods

1. Invoice requirements

In the earthwork contract, both parties should agree on the invoice requirements for payment. Clearly require the construction party to provide formal invoices to ensure legality and traceability.

2. Reason and method of deduction

When there is a breach of contract or quality issue, the owner has the right to deduct the corresponding amount. Therefore, it is necessary to clearly specify the conditions, reasons, and methods for breach of contract and deduction in the contract.

3. Penalty amount and enforcement method

If the construction party fails to complete the project according to the agreed time in the contract, the owner has the right to demand that the construction party pay a fine as agreed. The contract should specify the amount of fines and the method of enforcement.

第六节　课后练习

Ⅰ. Technical Terms Interpretation

1. substantial completion

2. final acceptance of project

3. quality supervisor

4. expert panel

5. technical staff

6. taking-over certificate

7. project initiation

8. dynamics of acceptance

9. implementation criteria

10. project audit

11. 交货付款

12. 货到付款

13. 装运付款

14. 现金结算

15. 拖期付款

16. 过期支票

17. 未兑现支票

18. 付讫支票

19. 财务审计报表

20. 资产负债表

Ⅱ. Sentences Interpretation

1. We've already prepared an acceptance plan.

2. We will finish items as soon as possible.

3. I'm the authorized representative of the acceptor. So please contact me when anything is needed.

4. The owner will not agree on acceptance until deliverables meet specific criterion in the agreement.

5. It is generally not reasonable to expect zero problem reports for a large system.

6. Please note this sum is paid a payment on the exgratia basis.

7. Excess payment shall be refunded and any deficiency repaid.

8. If payment is further delayed, we will be forced to take corresponding actions in accordance with the contract.

9. This is the estimated amount that we believe should be paid to us according to the contract provisions.

10. Please inform us of the name of the bank, the account name and number.

Ⅲ. Dialogue Interpretation

Mr. Smith: Well, we've settled the question of price, quality and quantity. Now what about the terms of payment?

Mr. Ma: 我们只接受不可撤销的、凭装运单据付款的信用证。

Mr. Smith: Could you make an exception and accept D/A or D/P?

Mr. Ma: 很抱歉,我们不能同意承兑交单付款方式。承兑交单只有在买卖双方非常了解、不会有拒付风险的情况下才使用。如果采用承兑交单付款方式,货已发出,到达目的地而被拒收,我们将面临极大的麻烦和资金损失。

Mr. Smith: How can that be? You ought to believe me, right? And to tell you the truth, a letter of credit would increase the cost of my import. When I open a letter of credit with a bank, I have to pay a deposit. That'll tie up my money and increase my cost.

Mr. Ma: 你和开证行商量一下,看他们能否把押金减少到最低限度。

Mr. Smith: Sorry, under such circumstances, D/A or D/P is the best payment method.

Ⅳ. Passage Interpretation

Further to our letter dated April 8, 2023 stating that if payment in full was not made within 7 days from the date of its receipt we would suspend our obligations, we have not received any (full) payment up to now. This is to inform you that, with immediate effect, we are suspending all our obligations under the contract.

The final certificate should have been issued on August 14, 2024. Two weeks have passed since that date and we have received no such certificate.

Without prejudice to our rights in this matter, if the final certificate is in our hands by September 14, 2024, we will take no further action on such breach.

Enclosed herewith the computations of the adjusted contract sum we made according to the new pricing method agreed by both sides at the meeting of November 10, 2024.

第十一章　工程项目索赔(参考答案)

第三节　对话口译

Dialogue 1: Negotiation on the Claims Caused by Flooding

E:尽管困难重重,但是该项目目前进展顺利。我谨代表我的公司对您的辛勤工作表示衷心的感谢。我希望我们能够共同努力,找到一个双方都能接受的解决方案。

C:To be honest, we are disappointed with your response. The two floods on June 12th and 24th caused us significant losses, as you can see from our contemporary record and claim calculations. I'm afraid if our claim is not settled, it may harm the smooth progress of the project.

E:在我来这里之前,我已经研究了所有关于你们要求补偿费用和延长工期的信件。由于高水位,你们遭受了损失,这是事实;损失的部分原因是你们不当的现场活动,这也是事实。我对所发生的事情感到遗憾。然而,你的要求太过分了。

C:I am confident that our requirements for cost and duration compensation are reasonable. Anyway, I would like to hear your opinion first.

E:让我直说吧。我们允许你们延长两个月的日程安排,并赔偿您总额为100000美元的损失。这是我们的最终报价。我相信这是公平和公正的。

C:Thank you for proposing this plan. We can accept a 2-month extension, but your cost compensation is significantly lower than our claim amount, which is based on our actual losses.

E:我不得不说,我们已经尽了最大努力。为了解决这个问题,双方都必须做出让步。

C:I admit that you have taken a step, but the pace is too small. We always hope to maintain a good working relationship with your side; We always hope to deliver successful projects to your side, but we also hope to receive fair and reasonable compensation. If necessary, we can resolve this claim through arbitration.

E:好吧,在这一点上,让我建议我们将赔偿金提高到150000美元。这个提议确实代表了我们的最终报价。如果你方不能接受报价,我们别无选择,只能将此事提交仲裁。

C:In the spirit of mutual understanding and facilitating our future cooperation, we accept your proposal, although this amount is far from enough to compensate for our losses.

E:我支持你关于相互理解与合作的呼吁。我将起草今天的协议作为合同的附录,并将其转发给您供您审查和签署。

第四节 篇章口译

Passage 1: A Speech on the Insights and Experience Summary of Construction Claims Management

Your excellencies and dear colleagues:

Hello everyone! In the process of project management, we often face various challenges and difficulties, one of which is claim management. Claims are an inevitable part of the project process, and we need to handle them with a positive attitude and professional methods. I would like to emphasize that claims are not necessarily a negative experience, but can be seen as an opportunity for learning and improvement in project management.

Firstly, we need to understand the essence of the claim. Claims are usually caused by unpredictable risks and changes in the project, which may include unclear contract terms, design changes, project delays, and other factors. Claim is not just a dispute or controversy, it is a legal requirement based on the contract, and an application for economic or time losses caused by unexpected situations in the engineering project. Understanding this can help us view claims more objectively, not only from a legal perspective, but also taking into account the underlying substantive issues. As a project manager, I will try my best to consider these factors in the project plan and contract to reduce the risk of claims. However, even with our best efforts, sometimes it is still impossible to avoid claims.

We should adopt a proactive attitude towards the handling of claims. Firstly, we need to carefully review the contract terms to ensure that we have the right to and exercise the relevant claims. At the same time, we also need to ensure that our claims are reasonable and sufficient, supported by sufficient documentation and evidence.

This includes but is not limited to contract documents, meeting minutes, design change notices, engineering logs, etc.

Before filing a claim, we need to have sufficient communication and negotiation with relevant parties. Communication is the key to solving problems. Through effective communication with stakeholders such as owners, designers, and contractors, we can better understand each other's needs and positions, and find the best way to solve problems. In the process of communication, we need to remain calm, rational, and avoid emotional expressions to ensure smooth and effective communication.

Additionally, we also need to pay attention to the timing of claims. Timely filing of claims can avoid further deterioration of the problem and help resolve it earlier. Before filing a claim, we need to conduct a thorough evaluation of the impact of the claim and possible solutions to ensure that our claim is reasonable and contributes to the smooth progress of the project.

Finally, during the claim handling process, we need to maintain good communication and collaboration with the team. The support and cooperation of the team are key to solving problems, and we need to encourage team members to provide suggestions and opinions, and work together to find the best solution. Through the efforts of the team, we can better cope with various challenges and ensure that the project can proceed smoothly as planned.

Overall, claim management is a complex and important aspect of project management. By adopting a proactive attitude, thorough preparation, and effective communication, we can better handle claims, learn from them, and improve the level of project management. Thank you all for your cooperation and support. Let's work together to ensure the successful completion of the project.

Thank you!

Passage 2: Response to Claims

您的索赔是关于泵房基础板中的预埋管道,施工图纸中的材料要求与合同文件中的技术规范不一致。您已经提交了几项技术澄清请求,并在 35 天后得到了答复。您认为最终决定太迟,在成本和工期方面造成了一些损失,有必要对延期和成本增加进行补偿。

但在我们看来,这种说法是无效的。在我们按照索赔程序详细审查索赔后,您会理解的。首先,您已在合同条款规定的时间内提交索赔。其次,您还在第一次指示后的 30 天内提交了索赔费用账单。但在我们根据合同条款中提到的程序进一步分析索赔事件后,您未能在 14 天内提交必要的信息。这意味着您无法为索赔提供足够的证据。

你在索赔中提出了明确的要求:第 1 点,与图纸进度表相比,施工图纸的出具和技术说明推迟了 42 天。为了赶上原计划,你调动了更多的材料、机器和人力,花费了 25397 美元。第 2 点,新图纸中的新材料价格比合同文件中的价格高出 15983 美元。但我们不能同意将您的描述作为成本分析。我们只能接受通过适当程序得出的结果。在这种情况下,我们不得不说,这种说法是不可行的,因此是无效的。

第六节 课后练习

Ⅰ. Technical Terms Interpretation

1. claims

2. penalty clauses

3. mediation

4. sample

5. differences and claims clause

6. defective packaging

7. delayed delivery

8. claims

9. claim letter

10. withdrawal of claim

11. 运费

12. 在途货物

13. 调查报告

14. 投诉

15. 提单

16. 执行职责

17. 次品

18. 货物状况良好

19. 由一方承担

20. 工艺

Ⅱ. Sentences Interpretation

1. In general, those transactions under which the importer does not receive goods of the kind, quality or quantity he expects, or the exporter does not receive the due payment lead to complaints and claims.

2. In the course of executing a contract, if one party breaks the contract and brings about economic losses to another party, the suffering party may ask the defaulting for compensation according to the contract stipulation.

3. Handling the suffering party's claim is called settlement of claim.

4. Sometimes when the loss is not serious, the party suffering the loss may not lodge a claim for compensation.

5. In case the goods delivered are inconsistent with the contract stipulations for quality, quantity, packaging, etc., the buyer should make a claim against the seller within the time limit of re-inspection and claim with the support of a survey report issued by a surveyor accepted by the seller.

6. In other words, the party that has failed to implement the contract must carry out his contract obligations in spite of his payment of the penalty.

7. When the exporter and the importer are involved in trouble, to settle the disputes it is advised that arbitration is better than litigation, and conciliation is better than arbitration.

8. We would refer to your consignment under the above order, which arrived here this morning.

9. Much to our regret, it was found upon examination that the goods are correct and in good condition except box No.7, whose box has been broken.

10. Therefore, the damages must have occurred in transit and we should not be held liable.

Ⅲ. Dialogue Interpretation

A:Hello, I am Mr. Zhang, the project manager. I would like to inquire about some issues related to our project regarding engineering claims.

B:你好,张先生。有什么具体的问题或者情况需要我解释吗?

A:Yes, we have noticed some project delays and additional expenses recently. I would like to know if it is possible to file a claim for the project.

B:哦,这确实是个重要的问题。可以具体告诉我有哪些方面导致了工程延误和额外费用吗?

A：Mainly during the construction process, inconsistencies were found in some design drawings, which led to the need for engineering adjustments and reconstruction. In addition, there have been issues with the supply of some materials, which have affected the progress of the project.

B：明白了。在这种情况下,我们可以考虑提出工程索赔。首先,我们需要收集相关的证据,包括设计图纸的不一致之处、工程延误的具体原因以及额外费用的明细。

A：Okay, we have started organizing the relevant files and data. However, before filing a claim, I would like to understand the process and possible outcomes of the claim.

B：索赔的流程通常包括提出索赔申请、相关方的审批和评估,最后是索赔的支付。在申请过程中,我们需要详细说明索赔的理由,并提供足够的证据来支持我们的主张。审批和评估的过程可能需要一些时间,但一旦获得批准,我们就可以得到相应的赔偿。

A：I understand. Thank you for your explanation. We will continue to prepare relevant materials and submit a claim application as soon as possible.

B：不客气,如果在准备过程中需要帮助,随时告诉我。希望一切顺利解决。

Ⅳ. Passage Interpretation

对于一般商品的适度合同,通常包括差异和索赔条款。如果交付的货物在质量、数量、包装等方面与合同规定不一致,买方应在重新检验的期限内向卖方提出索赔,并提供卖方认可的检验单位出具的检验报告作为支持。对于涉及大宗货物或高价值设备,如成套设备等的合同,还包括违约金条款,并在一方未能履行合同时,如未交货、延迟交货等。违约金通常占合同总价值的一定百分比,但支付违约金并不意味着可以撤销合同。换言之,未能履行合同的一方必须履行其合同义务,尽管支付了违约金。

第十二章 工程技术人员后勤保障(参考答案)

第三节 对话口译

Dialogue 1：Opening a Bank Account

银行职员：Good morning, welcome to the Bank of China. How can I help you?

Tim：早上好。我想开一个银行账户。

银行职员：Sure. What kind of account would you like to open? A current account or a deposit account?

Customer：有什么区别呢？

银行职员：A current account is a type of savings account. With it, you can handle cash deposits and withdrawals for personal transfer, exchange, inbound and outbound remittances. Current accounts can be used for internal transfers and exchanges between personal and other current accounts of the Bank of China. It can be used for domestic and international remittance and account entry. You can also activate online banking and mobile banking services. In addition, our bank will provide a certain amount of current interest. A fixed deposit account is a type of fixed deposit method for savings deposits. Cash deposits and withdrawals cannot be processed, and it cannot be used for transfers, incoming and outgoing remittances. We provide customers with fixed deposits of different maturities such as seven days, one month, three months, six months, nine months, and one year. After the maturity of the fixed deposit, it will be automatically transferred, and you can also withdraw the full amount in advance. Regular accounts can be opened for online banking and mobile banking services and can be operated independently on both online and mobile banking platforms; The prerequisite for opening a fixed-term account is to open a current or checking account. Compared to current deposits, time deposits have a higher interest rate.

Tim：明白了。我想申请开通信用账户。

银行职员：Sure, we will have you fill out an application form, please.

Tim：好的。

银行职员：How much of a credit limit were you looking for?

Tim：大概 10000 元的额度。

银行职员：Alright, we will see what we can do.

Tim：太好了。我刷卡后会有积分吗？

银行职员：Sure, with our general card, gold card, and titanium card products of Bank of China Series credit cards and Great Wall series credit cards, you will get 1 reward points for 1 yuan spent.

Tim：好的！申请表填好了，还需要什么吗？

银行职员：Yeah, your passport and work certificate please.

Tim：在这。

银行职员：Thank you, your account application is in process and your credit card will be mailed to your address within 15 to 20 business days.

Tim：谢谢！祝您今天过得愉快！

银行职员：Thank you. You too.

Dialogue 2：Office Building Renting

Estate Agent：下午好,史密斯房产中介。

陈东：Good afternoon. I am Chen Dong, office manager of the China Road Bridge Corporation. We plan to rent the four-storey office building on Cohen Street.

Estate Agent：您好,陈先生。您收到我前一天寄给您的租赁协议初稿了吗?

陈东：Yes. I read it and I am calling to discuss with you the provisions.

Estate Agent：好的。我来为您解释一下您需要了解的关键条款。

陈东：Thanks. Would you explain the obligations of the lessor in details?

Estate Agent：当然。首先,出租方应按期向承租人提供办公用房及其附属设施。

陈东：Please attach an appendix of furniture list in details in the formal lease agreement.

Estate Agent：好的。如果房屋及其附属设施因质量问题、自然灾害或其他灾害而损坏,出租方将支付翻新和维修费用。

陈东：I insist that the lessor offer the lessee an official invoice or a tax receipt on the same day of the receipt of the rent.

Estate Agent：这项条款就是您说的内容。

陈东：OK. What about the payment terms?

Estate Agent：承租人应在每个月的二十号前支付月租金。

陈东：That's all right. Is there a provision about the proper using of the premises?

Estate Agent：是的,有一条:承租人不得将房屋的全部或任何部分转租。

陈东：I see. What are the regulations for office decorating?

Estate Agent：经出租方批准,承租人可以装修房屋并增加新设施,但不得改变房屋的基本结构,不能影响房屋的正常使用。

陈东：Well, shall we move on and discuss the breach of contract?

…………

第四节　篇章口译

Passage：Why construction sites need proactive security？

从安全角度来看,建筑工地无疑是最具挑战性的需要受到保护的环境之一。与完全完工的建筑(工人和游客受到密切监控,并通过指定的出入口出入)不同,相比之下,建筑工地实际上来去自由,有多个出入口,不同的承包商和分包商可以随时出入。

正如大家所知的,建筑工地还汇集了很多非常受欢迎且易于转售的商品,从随着全美供应链危机而价格飙升的木材和铜,到电动工具和重型机械,不一而足。事实上,根据美国国家设备登记局和国家保险犯罪局发布的最新设备盗窃报告,仅2016年美国被盗设备的估计总价值就接近3亿美元。

一般情况下,承包商会采用各种策略来减少盗窃,例如安装周边围栏,雇佣警卫看管财产,并确保只有授权人员才能进入现场。然而,尽管巡逻警卫可能是预防和威慑潜在小偷的首选方法,但这样做也并非没有缺点。

首先,雇佣内部警卫或将职能外包给分包供应商可能成本高昂,尤其是在需要一人以上才能充分覆盖的大型场地。其次,警卫是人,无法全天候监视每个现场区域。自动机器人,或通常所说的无人地面车辆,也被认为是解决这一困境的一种方法。但是,它们的成本也很高,而且在充电后只能在一段有限的时间巡逻。

视频监控角色的变化

一直以来,视频监控在保护建筑工地方面也发挥着重要作用。然而,即使摄像机放置和维护得当——这在建筑区域内是一个重大挑战——拍摄的视频通常也只用于记录已经发生的犯罪或伤害。此外,人们已经习惯了摄像头的无处不在,以至于它们可能无法发挥曾经的威慑作用。

以机器学习和人工智能技术为动力的新视频分析工具的开发可能有助于使监控成为建筑工地更积极主动的安全工具。例如,户外安全应用程序历史上最大的挑战之一是,动物或树叶等物体在风中吹过时,会意外激活房屋周围的虚拟绊网。然而,视频分析可以学习并区分好奇的鹿或倒下的树枝与进入该地产的人或车辆。一些视频分析还可以在人员、车辆和物体穿过场景时对其进行分类和跟踪,为安全和急救人员提供对潜在入侵者的准确描述。

从被动安全防卫过渡到主动安全防卫

即使有了视频分析,仅靠监控也无法创建主动的安全策略。将监控和分析与现场操作员相结合,有助于建筑安全专业人员管理安全威胁。

有警卫在值班,并不意味着他们可以对每一次潜在的入侵做出反应,即使有提前通知的帮助。小偷只需很短的时间就可以攻破一栋正在开发的建筑,将价值数千美元的线束、木材和工具堆放在一辆皮卡车的车厢里,或者更糟的是,在当前供应链紧张的情况下,他们带着重型设备离开,这些设备已经不像以前那么容易获得了。

某天,如果有人进入房屋的视频被捕捉到,24 小时监控中心的现场工作人员可以收到入侵警报,并采取一系列行动来解决这种情况,例如直接警告入侵者,他们正在被观察和记录,并通知当局。在那些还雇了警卫的建筑工地上,工作人员可以将他们派往入侵的确切区域,并让他们接手对应工作。

多年来,视频监控主要作为事件后的调查工具。尽管如此,分析技术的出现和实时监控的发展意味着有更多的策略可以解决当今建筑工地上存在的无数风险。从被动的安全态势转变为积极主动的安全态势,可能意味着一个项目能否按时、按预算完成,与一个项目是否延误、财务超支的区别。

第六节　课后练习

Ⅰ. Technical Terms Interpretation

1. logistics support

2. warehouse

3. goods and materials supply

4. fixed assets

5. equipment maintenance

6. financial plan

7. accounting

8. vehicle management

9. water and electricity management

10. healthcare management

11. 后勤管理部门

12. 国有资产

13. 能源管理

14. 基础设施建设

15. 维修改造

16. 社会服务企业

17. 绿化养护

18. 物业服务

19. 水电暖供应

20. 宿舍管理

21. 医疗卫生

II. Sentences Interpretation

1. Getting vaccinated often depends on whether or not it's convenient to do so. To help make vaccination as easy as possible, employers can pay for childcare or the travel costs of getting to vaccination facilities, or offer paid time off for their employees to get vaccinated. It's important to give employees the time and flexibility they need to attend vaccination appointments, and even recover from vaccination.

2. One of the best ways to address people's concerns about getting vaccinated is to refer them to someone they trust. Remember that these concerns can be emotional for people and touch on issues outside of science, such as their personal experiences and their perceptions of poor or unfair treatment in the past. Don't overwhelm people with facts and information. Instead, acknowledge their perspectives or experiences and speak to their motivations, not what you think they need to hear.

3. The facilities and equipment for property management mainly include power supply lines, various distribution equipment, water supply and drainage pipelines, secondary domestic water pressurization equipment, elevators, fire protection equipment, etc., ranging from a ubiquitous street lamp in the dormitory area to hundreds of thousands of power distribution devices.

4. Faults in internal and external power supply systems, sudden power outages in dormitories, emergency power supply system activation management, and timely repair of power line faults, as well as emergency repair of water supply pipelines are all work of logistics support department. Regardless of whether these malfunctions occur during the day or late at night, in severe cold or heat, the logistics support department is dutifully committed to immediately engaging in emergency repairs, with the aim of restoring and ensuring the normal living order of residents as soon as possible.

III. Dialogue Interpretation

Peiling: Vanita! How's it going?

Vanita: 嘿, 佩玲, 你已经预约了流感疫苗注射了吗?

Peiling：Yes, I just did! I'm scheduled for 10 am. How about you?

Vanita：我预约的是10：30。我们的预约时间很接近,你想在诊所碰面,一起去吗?

Peiling：It sounds good. Let's meet outside the clinic at 9：45 am?

Vanita：没问题,我们诊所见。

Peiling：Since our appointment times are slightly different, do you think they'd allow us to get our shots together?

Vanita：啊,我也不确定,我们问问。

Vanita：你好,我朋友的预约时间是10点,我的预约时间是10：30。请问我们可以一起注射流感疫苗吗?

Nurse：我查询一下。可以告知我你们的姓名吗?

Vanita：我叫Vanita,我朋友叫佩玲。

Nurse：你们可以一起注射。现在前面还有一位,然后我们将为你们接种疫苗。

Vanita：谢谢!

Nurse：准备就绪了,谁先打呢?

Peiling：I'll go first since my appointment was earlier.

Nurse：您想打哪一只手臂,左手还是右手?

Peiling：Left arm, please.

Nurse：您打哪只手?

Vanita：右手。

Nurse：好了! 我给你们贴上止血贴。

Ⅳ. Passage Interpretation

如果薪资和奖金的谈话进展不顺利,经理们应该花更多的时间倾听员工的意见,以了解他们来自哪里以及他们的担忧是什么。

当涉及薪酬的谈话陷入僵局时,更有"好奇心"的经理们往往会收获颇丰。例如,他们可能会了解到,员工可能认为他们的工作未能达到竞争对手或市场上类似工作的同等待遇。这可能是真的,也可能不是真的,在做出最终决定之前,你最好进行深入调查。

通常情况下,围绕工资和奖金的对话后,需要召开一次后续会议,让经理们了解更多的事实,并确保与员工的联系取得积极的结果。双方的目标是达成一个让双方都满意的决定,让你的员工感受到公司的重视和适当的补偿,而你对各类数据以及企业的收入和产出平衡也感到满意。

　　你应该对薪资和奖金相关的对话加以重视，因为它们会对你的企业生产力产生巨大影响——无论是你的团队感到被低估、失去动力或决定离开，还是你的员工对他们的薪水感到高兴，他们对公司的宣传都得到了提升。把时间花在有效的谈话上，你的员工很快就会向所有相关人员赞美你，帮助你吸引新的有才华的团队成员和新客户。

参 考 文 献

［1］陈安定.英汉比较与翻译［M］.北京:中国对外翻译出版公司,1998.

［2］陈焕江,徐双应.交通运输专业英语［M］.2 版.北京:机械工业出版社, 2008.

［3］陈俊,常保光.建筑工程项目管理［M］.北京:北京理工大学出版社, 2009.

［4］陈生保.英汉翻译津指［M］.北京:中国对外翻译出版公司,1998.

［5］陈新元.工程项目管理:FIDIC 施工合同条件与应用案例［M］.北京:中国水利水电出版社,2009.

［6］陈勇强,张水波.国际工程索赔［M］.北京:中国建筑工业出版社,2008.

［7］陈振东,黄樱.口译中的模糊信息处理［J］.上海科技翻译,2004(1): 36 - 39.

［8］崔军.FIDIC 分包合同原理与实务［M］.北京:机械工业出版社,2010.

［9］戴文进.科技英语翻译理论与技巧［M］.上海:上海外语教育出版社, 2003.

［10］范文祥.英文合同阅读与分析技巧［M］.北京:法律出版社,2007.

［11］冯树鉴.实用英汉翻译技巧［M］.上海:同济大学出版社,1995.

［12］葛亚军.英文合同［M］.天津:天津科技翻译出版公司,2008.

［13］GOULD F E,JOYCE N E.工程项目管理:英文版［M］.北京:中国建筑工业出版社,2006.

［14］顾慰慈.工程项目质量管理［M］.北京:机械工业出版社,2009.

［15］国际咨询工程师联合会.设计采购施工(EPC)/交钥匙合同工程条件［M］.唐萍,张瑞杰,等译.北京:机械工业出版社,2021.

［16］国际咨询工程师联合会.生产设备和设计—施工合同条件［M］.唐萍,张瑞杰,等译.北京:机械工业出版社,2021.

［17］何伯森.工程项目管理的国际惯例［M］.北京:中国建筑工业出版社,

2007.

　　[18]何伯森.国际工程承包[M].2 版.北京:中国建筑工业出版社,2007.

　　[19]黄坚,谷丰.新纬度通用工程英语听说教程[M].北京:商务印书馆,2019.

　　[20]贾荣香.涉外工程英语[M].北京:对外经济贸易大学出版社,2013.

　　[21]郎守廉.国际工程实务英语百事通:词汇·短语·例句·范例[M].北京:中国建筑工业出版社,2011.

　　[22]厦门大学口译教研小组.口译教程[M].上海:上海外语教育出版社,2006.

　　[23]李天舒.最新简明英语口译教程[M].西安:世界图书出版西安公司,2001.

　　[24]林超伦.实战口译　学习用书[M].北京:外语教学与研究出版社,2004.

　　[25]刘和平.口译技巧:思维科学与口译推理教学法[M].北京:中国对外翻译出版公司,2001.

　　[26]刘和平.口译理论与教学[M].北京:中国对外翻译出版公司,2005.

　　[27]吕文学.国际工程项目管理[M].2 版.北京:科学出版社,2021.

　　[28]梅德明.通用口译教程[M].北京:北京大学出版社,2007.

　　[29]梅德明.英语口译教程[M].北京:高等教育出版社,2003.

　　[30]欧阳倩华.基础口译[M].上海:上海交通大学出版社,2020.

　　[31]史澎海.工程英语翻译[M].西安:陕西师范大学出版社,2011.

　　[32]孙建光,李梓.工程技术英语翻译教程[M].南京:南京大学出版社,2021.

　　[33]孙丽丽."一带一路"建设与国际工程项目管理[M].北京:中国建筑工业出版社,2023.

　　[34]孙有中,查明建,张爱玲.理解当代中国:高级汉英口译教程[M].北京:外语教学与研究出版社,2022.

　　[35]邬姝丽.实用英语高级口译教程[M].北京:外语教学与研究出版社,2009.

　　[36]吴之昕,刘鸣笛.最新国际工程承包情景会话 20 讲[M].北京:中国建

筑工业出版社,2021.

　　[37]吴之昕,张珺.工程承包实用英语会话[M].北京:中国建筑工业出版社,2003.

　　[38]希基.语用学与翻译[M].上海:上海外语教育出版社,2001.

　　[39]邢厚媛.带资承包及 BOT 发展走势:中国企业开展带资承包面临的问题及对策[J].施工企业管理,2006(6):8-9.

　　[40]许建平.英汉互译入门教程[M].2 版.北京:清华大学出版社,2015.

　　[41]俞戊孙.建筑施工实用英语会话[M].北京:中国建筑工业出版社,1999.

　　[42]岳峰,曾水波.科技翻译教程[M].北京:北京大学出版社,2022.

　　[43]张水波,刘英.国际工程管理实用英语口语:承包工程在国外[M].北京:中国建筑工业出版社,1997.

　　[44]张文,韩常慧.口译理论研究[M].北京:科学出版社,2006.

　　[45]邹霞,刘黎.工程英语口译实务[M].北京:人民交通出版社股份有限公司,2015.

　　[46]BALLARD G,HOWELL G. Shielding production:an essential step in production control[J]. Journal of construction engineering and management,1998,124 (1):11-17.

　　[47]CHAROENNGAM C,SRIPRASERT E. Assessment of cost control systems:a case study of Thai construction organizations[J]. Engineering, construction and architectural management,2001,8 (5/6):368-380.

　　[48]FLYVBJERG B. The Oxford handbook of mega project management[M]. Oxford:Oxford University Press,2017.

　　[49]GENTZLER E. Contemporary translation theories [M]. London: Routledge,1993.

　　[50]HOFFMAN A W. From heresy to dogma:an institutional history of corporate environmentalism[M]. San Francisco:New Lexington Press,1997.

　　[51]HORMAN M J,THOMAS H R. Role of inventory buffers in construction labor performance[J]. Journal of construction engineering and management,2005, 131 (7):834-843.

[52] NEWMARK P. Approaches to translation [M]. London: Pergamon Press Ltd. ,1982.

[53] NIDA E A. Language, culture, and translating [M]. Shanghai: Shanghai Foreign Language Education Press,1993.

[54] NIDA E A. Toward a science of translating [M]. Leiden: E. J. Brill,1964.

[55] NIDA E A, WEARD J. From one language to another [M]. London: Thomas Nelson Publishers,1986.

[56] SCHULTE R, BIGUENET J. Theories of translation [M]. Chicago: The University of Chicago Press,1992.

[57] SPIELMANN R, BRISLIN R W. Translation application and research [M]. New York: Gardner Press,1992.

[58] WANG Q Y, HAN X R. Selected readings in hydrogeology [M]. Shanghai: Shanghai Foreign Language Education Press,2006.

[59] WILSS W. The science of translation [M]. Shanghai: Shanghai Foreign Language Education Press,2001.